I0475577

Arguing with the DSM-5:
Reflections from the Perspective of Social Work

Arguing with the DSM-5:
Reflections from the Perspective of Social Work

Nora L. Ishibashi, Ph.D., Editor

© 2014 Nora L. Ishibashi, Ph.D.
405 North Wabash Avenue, Suite 1303
Chicago, IL 60611

All rights reserved. This book or any portion thereof may not be reproduced or used in any manner whatsoever without the express written permission of the publisher except for the use of brief quotations in a book review or scholarly journal.

First Printing: 2014

ISBN: 978-1-312-44751-6

Ordering information:

This book may be ordered from http://www.lulu.com

Dedication

This book is dedicated to all the social work students who have committed themselves to lives of care and responsibility.

Contents

Preface

Nora L. Ishibashi, Ph.D.

The *Diagnostic and Statistical Manual of Mental Disorders* (DSM), which is published by the American Psychiatric Association, is the standard catalog used by most mental health practitioners in the United States to provide diagnoses for psychological distress. Like earlier versions of the DSM, the current edition, which is the fifth complete revision, continues to create controversy as experts argue the merits and pitfalls of its definitions and implications.

This book collects papers from first year master's degree students in the Social Work School at Loyola University of Chicago. Within the context of their class in Human Behavior in the Social Environment, social work students study the DSM as one of the many ways social workers assess the needs of their clients. These papers represent the students' responses to the diagnostic categories and their arguments both supporting and opposing the formulations found there.

Psychological distress is difficult to assess because most of the time we have only the report of the client together with our own observations as evidence for identifying the problem. Unlike medical problems, psychological problems cannot be detected using blood tests, EEG's, blood pressure monitors, or other mechanical measures. We are left with our subjective judgment and the client's description of his or her subjective experience as means of understanding what may be happening. As a result, the meaning of that experience can be interpreted and described in many different ways, which means naming the problem can be said to identify a reality but can also be said to create a reality.

Engaging with another person to address problems in his or her life is a serious charge, which can have lasting effects. That is why social work can be so helpful when a client is facing stressors of any kind, and it is also why social work has great power. With great power comes great responsibility, and the social work profession has over its long tradition generated many thoughtful books, much careful research, and countless knowledge building conferences, meetings, and curriculums. We have taken our responsibilities seriously and developed ethical standards and caregiving approaches with the client's well-being as our central interest.

The papers collected here are a contribution to that body of knowledge, containing the perspectives of the newest members of the social work profession. Students come to social work with good intentions, kind hearts and curious minds. It is a great privilege to be involved in the preparation of the next generation of caregivers who face the complex and serious problems of our current culture. We had many deep discussions regarding assessment as students developed their own ideas and opinions. You will find these papers to be thought-provoking and nuanced. They are the voices of the next generation of social workers.

Alcoholism is Not a Disease

Morgan Albrecht

In the Fifth Edition of the Diagnostic and Statistical Manual of Mental Disorders (DSM-5), the section on Substance-Related and Addictive Disorders has 10 different classes of drugs. The essential feature of a substance use disorder is a cluster of cognitive, behavioral, and physiological symptoms indicating that the individual continues using the substance despite significant substance-related problems (American Psychiatric Association [APA], 2013). A more specific focus states "alcohol use disorder is defined by a cluster of behavioral and physical symptoms, which can include withdrawal, tolerance, and craving" (APA, 2013).

However, since the DSM-5 uses more of a medical model it portrays alcohol use disorder as a fatal and incurable illness requiring medical attention. Using this model to provide the majority of the diagnostic criteria, the DSM-5 fails to take into consideration important social and environmental factors. Furthermore, it neglects to take into account the individual in relation to a social context and its role in the diagnosis of alcohol use disorder, which ultimately can then lead to over-diagnosing. Those diagnosed with alcohol use disorder or alcoholism are now forced to take on a new identity as an alcoholic leaving them to deal with the stigma associated with the term, and they no longer view themselves as "normal".

This leaves us with several questions. If an individual displays what the DSM-5 labels as mild symptoms, such as drinking more than intended or an increase in tolerance, does that automatically lead to being referred to as having an addiction (APA, 2013)? Also, if we neglect to look at the complexity of the experiences and the individual happens to meet the mild criterion in their first several times using a substance, are they automatically put into the category used for hard core users? We need to look at the multiple systems that encompass an individual when diagnosing along with looking at the context of the use.

Social Learning Theory

Albert Bandura first addressed social learning theory (SLT) in that social behavior is learned through observing, modeling and imitating the actions of others in various interactions. When looking at alcohol use, "social learning theory suggests that drinking, as a social behavior, is acquired and maintained by modeling, social reinforcement, and the anticipated effects of alcohol" (Abrams, Niaura, 1987, p. 136). Alcohol consumption behaviors are formed through social influences and the social environment which includes peers, family, groups, organizations, the community and it even includes one's culture. Therefore, it is important to look at these reciprocal relationships and take into account any correlation between individual alcohol use and the corresponding interactions within the social environment.

Social Modeling and Drinking Behaviors

If an individual decides to pursue a higher education such as college, they most likely will have the opportunity to join a Greek organization (that is, a fraternity or sorority). These affiliations provide an environment that is known for heavy alcohol use among peers along with stronger drinking norms. When looking at the social group with STL, individual behavior is shaped by the social environment hence members of Greek organizations engage in greater alcohol consumption. This is especially true if the individual observes the organization members as having a positive regard towards alcohol consumption making it easier to rationalize their own consumption. Also, there may be an intrinsic reinforcement wherein the individual feels an internal reward for modeling drinking behaviors in hopes of feeling a sense of pride and accomplishment in being accepted into a Greek organization of their choice.

An early aspect of shaping the norms of drinking behavior is during a ritual known as rush. At this time individuals interested in joining a Greek organization, otherwise known as potential members, get to know each other through a series of social gatherings and social events. Alcohol use can be used to enhance these social interactions in order to facilitate bonding. Whether or not college students decide to join a Greek organization, college is known as an important period of time in a student's life where socialization with their peers might include binge drinking. Studies consistently document that the highest rates of heavy alcohol use and alcohol use disorders occur in the college-age population, with higher rates among college students than among their noncollege peers (Dawson, Grant, Stintson, & Chou, 2004). The National Institute of Alcohol Abuse and Alcoholism (NIAAA) defines binge drinking as a pattern of drinking that brings blood alcohol concentration (BAC) levels to 0.08 g/dl. This usually occurs after 4 drinks for women and 5 drinks for men-in about 2 hours (http://www.niaaa.nih.gov/alcohol-health/overview-alcohol-consumption/moderate-binge-drinking).

One of the criteria for alcohol use disorder is consuming larger amounts of alcohol than intended. Therefore, this particular diagnostic criterion would be met because the large amount of alcohol consumed during binge drinking corresponds with the social norms of a lot of college drinking habits. Another criterion of a mild diagnosis of alcohol use disorder is developing a tolerance. A college lifestyle that includes binge drinking on the weekends leads to the possibility of building a tolerance to alcohol and would then be consistent with the diagnosis of mild alcohol use disorder according to the DSM-5.

In addition, according to a recent publication in the *Psychology of Addictive behaviors*, young adulthood from age 21 to the early 30s is an important developmental period for alcohol use. For most young adults, consumption of alcohol peaks in the early 20s and then declines as many complete college and/or begin families and careers (Kosterman et al., 201). Therefore, not only does the social aspect of the college lifestyle but also the developmental period during this time period need to be taken into consideration before making any diagnosis. Once the individual graduates, begins his or her career and possibly starts a family, the majority of these individuals disengage from the behaviors displayed during their college years. If the social aspect of the drinking behavior is not taken into account, a misdiagnosis is something that an individual will carry with him or her into proceeding developmental periods across the lifespan.

Family Modeling and Drinking Behaviors

One of the most influential systems in understanding human behavior is family systems, and they provide an instrumental part of the social organizations in STL. The 2012 annual survey sponsored by the Substance Abuse and Mental Health Services Administration (SAMSHA) found an annual average of 7.5 million children younger than the age of 18 live with a parent who has an alcohol use disorder and they are four times more likely than other children to develop alcohol problems themselves (http://www.samhsa.gov/data/). Continuing to look at alcoholism as a disease one would expect the offspring of alcoholic parents would carry a genetic disposition and inherit the disease in adulthood. However, looking at alcoholism through a social lens, family modeling and differential reinforcements can help in explaining why some individuals become alcoholics and others do not even though they are immersed in similar environments.

Several factors appear to influence an individual's drinking, and utilizing STL, one would look at the interactions between parents and the child. Modeling plays a role because the drinking pattern displayed determines what the child learns about alcohol use. Vicarious learning (modeling) is a form of learning where "people acquire new knowledge and behavior by observing other people and events, without engaging in the behavior themselves and without any direct consequences to themselves" (Wilson, 1988, p. 240). The article *Parenting Influences on the Development of Alcohol Abuse and Dependence* presents the notion that "children's perceptions of parental drinking quality and circumstance appear to influence their own drinking frequency" (p. 205). If a parent's drinking pattern is one of excess leading to consequences such as family aggression, verbal abuse, parental absence or behaviors that result in incarceration, the consequences modeled are extremely negative. Using STL one would look at the reciprocal relationship between the parent and child. Clearly, the parent's behaviors and consequences around alcohol consumption would influence the development of the child's future alcohol use behaviors such as choosing not to consume alcohol out of a fear of having the same negative behaviors and/or consequences modeled by their parents. Elevated rates of alcohol exposure shape a child's understanding of drinking and reinforce the expectancies of alcohol consumption.

Free Will and Choice

By continuing to look at addiction as a disease and assuming that certain individuals have a genetic predisposition to alcohol, we eliminate the concept of free will. Specifically this eliminates the concept of the freedom of choice that an individual has in deciding whether or not to consume alcohol. A disease model of addiction focuses on biological and genetic factors, meaning that some sort of medical treatment would be required. However, just because it has been said that alcoholism is a disease does not necessarily make it true. There has yet to be concrete evidence supporting this notion.

Despite billions of dollars in scientific research and decades of blaming one's biology, there has never been an alcoholic gene discovered nor has there ever been proved a biological flaw that forced someone to pick up the first drink of alcohol and continue drinking into an out of control state (Burras, 2011, p. 68).

This disease concept is one way to persuade those diagnosed with alcohol use disorder and labeled alcoholics that the only solution is through medical attention, and this notion continues to financially feed the treatment center industry and pharmaceutical companies. A major system that supports the disease model of addiction and utilizes this concept is the 12-step recovery program Alcoholics Anonymous (AA). According to their website, "Alcoholics Anonymous is an international fellowship of men and women who have a drinking problem" and the first step of the program states "we admitted we were powerless over alcohol" (http://www.aa.org/pages/en_US/what-is-aa). Using the word powerless implies that the alcoholic lacks control over his or her drinking therefore removing responsibility over their actions. Powerlessness can also be seen as being helpless and reinforces the idea that the alcoholic is a victim to biological and genetic forces that lead him or her to drink excessively. It also removes the notion of freedom of choice. However, in retrospect these individuals do not have someone physically pouring alcohol down their throats and forcing them to drink. Even if they are unable to physically remove themselves from a location where alcohol is present, they still have the choice whether or not to consume it.

Concluding Thoughts

There is no concrete answer to whether or not alcohol use disorder should be included in the DSM-5. Until research can support the medical concept of alcoholism being a disease by showing a biological or genetic basis behind it, one must question this diagnosis. Currently the DSM-5 lacks significant evidence-based findings and presents the alcohol use disorder diagnosis utilizing criteria based on a cluster of symptoms disregarding social and environmental factors. This results in medical health practitioners or clinicians basing their diagnosis on their discretion without utilizing all the variables that are included in an individual's life.

Even having the DSM-5 as a clinical resource in that it provides diagnostic criteria and even with the newest release of the DSM-5 which now provides a broader dimension of symptoms including mild, moderate and severe, the question remains how to ensure that the symptoms are being accurately portrayed in order to provide the correct diagnosis. Screening is a tool used to assist mental health practitioners and clinicians. When using this tool, an individual is required to answer a variety of questions geared towards measuring his or her alcohol consumption. However, by using a questionnaire the clinician trusts the individual to provide accurate and honest responses. There is always the possibility that an individual will lie due to the fear of being labeled an alcoholic. Another obstacle is that the screening tool selected must be tailored to the population tested while taking into consideration diversity such as a wide range of ethnic and cultural backgrounds.

When looking at alcohol use disorder using STL, one must observe an individual's behavior by looking at reciprocal relationships and environmental factors. Alcohol use disorder is a complex disorder, and there is not a linear relationship between alcohol consumption and the severity of use or a relationship between alcohol use disorder and experimentation such as that found in a college setting. Even with the argument that alcohol use disorder cannot be diagnosed utilizing a book that heavily relies on a medical model of addiction, it is unrealistic to remove it altogether from the DSM-5. It is more feasible to include it in a behavioral addictions category along with gambling disorder or caffeine-related disorders and tobacco-related disorders and group them together in their own section. The DSM-5 clearly states the behavioral addictions "activate reward systems similar to those activated by drugs of abuse and produce some behavioral symptoms that appear comparable to those produced by the substance use disorders" (APA, 2013). Even more so it would be beneficial for clinicians to utilize The International Classifications of Diseases (ICD) instead of first going to the DSM-5 when presented with symptoms of substance-related disorders or utilize them together. The World Health Organization and *The ICD-10 Classification of Mental and Behavioural Disorders: Clinical Descriptions and Diagnostic Guidelines* states:

> Narrowing of the personal repertoire of patterns of psychoactive substance use has also been described as a characteristic feature (e.g. a tendency to drink alcoholic drinks in the same way on the weekdays and weekends, regardless of social constraints that determine appropriate drinking behavior) (p. 5).

By doing so, clinicians would include the characteristics around the drinking behaviors to determine if they constitute a substance abuse diagnosis. All in all, one must utilize all available resources before diagnosing an individual as an alcoholic; otherwise they are left to carry their own Scarlet Letter for the remainder of their lives.

References

Abrams, D. B., Niaura, R. N. (1987). Social Learning Theory. In A. Blane, H. T. & Leonard, K. E. (Ed. 1), *Psychological Theories of Drinking and Alcoholism* (pp. 131-178). New York, NY: The Guilford Press.

American Psychiatric Association. (2013). *Diagnostic and statistical manual of mental disorders* (5th ed.). Arlington, VA: American Psychiatric Publishing. Burras, J. (2011). Alcoholism Is Not a Disease. In A. Watkins, E. *Alcohol Abuse* (pp. 67-81). Farmington Hills, MI: Greenhaven Press.

Dawson, D. A., Grant, B. F., Stinson, F. S., & Chou, P. S. (2004). Another Look at Heavy Episodic Drinking and Alcohol Use Disorders among College and Non-college Youth. *Journal of Studies on Alcohol, 65,* 4. Retrieved from http://luclibrary.worldcat.org.flagship.luc.edu/title/another-look-at-heavy-episodic-drinking-and-alcohol-use-disorders-among-college-and-noncollege-youth/oclc/110240117

Jacob, T., & Johnson, S. (1997). Parenting Influences on the Development of Alcohol Abuse and Dependence. *Alcohol Health & Research World, 21,* 3. Retrieved from http://luclibrary.worldcat.org/oclc/4598925143&referer=brief_results

Kosterman, R., Hill, K. G., Lee, J. O., Meacham, M. C., Abbott, R. D., Catalano, R. F., & Hawkins, J. D. (2014). Young Adult Social Development as a Mediator of Alcohol Use Disorder Symptoms From Age 21 to 30. *Psychology of Addictive Behaviors: Journal of the Society of Psychologists in Addictive Behaviors, 28,* 348-358. doi: 10.1037/a0034970

Wilson, G. T. (1988). Alcohol Use and Abuse: A Social Learning Analysis. Theories on Alcoholism. In. A. Chaudron, C. D. & Wilkinson, D. A. *Theories on Alcoholism* (pp. 239-287). Toronto, ON: Addiction Research Foundation.

World Health Organization. (2013). *The ICD-10 Classification of Mental and Behavioural*

Disorders: Clinical descriptions and diagnostic guidelines. [PDF file] Retrieved from http://www.who.int/substance_abuse/terminology/ICD10ClinicalDiagnosis.pdf

Should Narcissistic Personality Disorder Be In The DSM-5?

Laura Colón

Introduction

In today's modern, high-tech world, narcissism is constantly present. Online social media has made it socially acceptable to constantly share opinions, achievements, and stories that would have previously been unsolicited by peers. Now, with the click of a button, people can boast about themselves and receive immediate feedback from their friends, family, colleagues, and people they may have never even met. In the past, people speaking of their success could just be cited as having high self-esteem, but in today's media-driven society, does this explanation still stand? Social constructionism assumes that one's worldview, worth, and sense-of-self are developed not by the individual, but in accordance to others in society. With the rise in "reality" television, should we be viewing these families and figures as an accurate portrayal of today's society? Looking through the lens of social constructionism, we will review the diagnosis of narcissistic personality disorder (NPD) and modern society to make a judgment of its placement in the DSM-5.

Narcissistic Personality Disorder: The Diagnosis

The DSM-5 states that personality disorders are an enduring pattern of inner experience and behavior that deviates markedly from the expectations of the individual's culture, is pervasive and inflexible, has an onset in adolescence or early adulthood, is stable over time, and leads to distress or impairment (APA, 2013). There are three categories of personality disorders, clusters A, B, and C. Narcissistic Personality Disorder is a Cluster B personality disorder meaning that individuals appear dramatic, emotional, or erratic. This cluster includes antisocial, borderline, histrionic, and narcissistic personality disorders. According to the DSM-5 a Narcissistic Personality Disorder is described as a pattern of grandiosity (in fantasy or behavior), need for admiration, and lack of empathy, beginning in early adulthood. Elsa Ronningstam put it best, "few personalities have lent themselves to the imagination of subtypes and variations as well as has the narcissistic personality, and few have been observed and documented with such fascinations, puzzlement, and awe (2005, p.7)."

For one to be diagnosed with a narcissistic personality disorder, one must fulfill five of the nine conditions, as recorded in the DSM-5. The first, and most prominent and defining feature is to have a feeling of grandiose self-importance. The patient may often embellish accomplishments and abilities and/or anticipate recognition as superior without actually achieving anything. Second, one must be preoccupied with fantasies of unlimited success, power, brilliance, beauty, or ideal love. For example, the patient may be obsessed with the idea of becoming a billionaire, but may work at an hourly wage and have no real drive or potential to actually earn such a large amount of money. Third, one may believe he or she is extremely unique and can only be understood by, or associate with those of such high stature. This person may think that he or she is so fantastic, he or she is only worthy of being around A-list celebrities, business leaders, and others of such prominent social status. A fourth criterion in diagnosis, is that the patient requires extreme and constant admiration from those around him or her. The fifth criterion, is quite similar to the two previous criteria in that the patient has a strong sense of entitlement, with unreasonable expectations of special treatment. This relates to the "diva complex," many of today's prominent celebrities take on as they become successful.

The last four criteria describe how the diagnosed treats others, versus how they feel about themeselves. Another potential condition in NPD is that the patient is interpersonally exploitative, meaning they take advantage of the people in their life to accomplish their own goals. Seventh, the patient lacks empathy. They are unable to sympathize with their peers and are unwilling to recognize the feelings and needs of others. The eighth circumstance describes that the patient is most likely frequently jealous of other peers, especially if they outshine them. In turn, they are often certain that others must be envious of them. The last condition in narcissistic personality disorder is that the client shows an egotistical and proud attitude, with accompanying actions. Again, if one fulfills five of the nine criterions listed above, they can be diagnosed with NPD, according to the DSM-5.

There are a few other interesting things to note while studying NPD. The diagnosis is often associated with anorexia nervosa. It is also frequently associated with substance abuse, particularly cocaine—an interesting connection as a side effect of cocaine-use is a feeling of supremacy. Studies have also shown that 50% - 75% of those diagnosed with NPD are male (APA, 2013). Diagnosed individuals often have difficulty adjusting to the aging process. Between their dependency on others and failing features (looks, intelligence, mobility), it is common for those with NPD to approach aging with a negative outlook. Lastly, one of the more notable features of NPD is the inability to love others. Even in some severe psychotic disorders, object love may remain relatively undisturbed while a profound disturbance in the realm of narcissism is never absent (Kohut, 2009, p.8).

Social Constructionism: A Theory of Human Behavior

The focus of social constructionism is how sociocultural and historical circumstances shape individuals and the formation of knowledge. In other words, how do individuals create themselves within their society? From this standpoint, individuals can be in charge of their own lives and be free to be self-guided and self-governed in their pursuit of personal development (Sutherland, 2013,

p. 375). Social constructivists believe all of life's experiences are personal and individuals recreate themselves through a constantly moving and changing process. Knowledge is created through a multitude of societal and historical interactions, which is why the idea of one's self is never static. Postmodern theorists believe social interaction is grounded in contexts of language, culture, and history. There is much skepticism in the therapist being the expert and the client is seen as being the most knowledgeable in their life story. Therefore, all sensations must be approached with doubt in order to understand how the client constructs their reality. Another common thought is the belief that humans are self-interpreting beings. Again, this leads back to the thought that the client is the expert in his or her own life.

In approaching the phases of helping, therapists generally adhere to the thought that knowledge and meaning are created linguistically though dialogue and through social institutions constituted in part though interpersonal negotiations. In the therapeutic relationship, this meaning is created through interaction between therapist and client and is intertwined with context (Coady et al., 2008, p.184).

Narcissistic Personality Disorder through a Social Constructionism Lens

Social constructionism emphasizes the power of social interaction and culturally shared assumptions for shaping knowledge and meaning. Of particular interest is the intersection among power, social discourse, and culture, which is seen as an engendering meaning for people (Coady et al., 2008, p. 56). Social interaction has changed drastically in the past twenty years in our society. In the past, a group of friends from high school who moved far away from one another would keep in touch through writing letters and/or regular telephone conversations. Now, these long-distance friendships can be maintained in the online realm. Thanks to online social media posts, a former classmate can gain access to loads of personal details in a peer's life without ever having to communicate with them. At their twenty-year high school reunion, the question is not, "how have you been?" but "how's your leg after that surgery you had a few months ago? I read about that!" No one is pulling out photographs from their wallets to show off their children to their friends who may have not known about said children. Instead, one is more likely to ask about how the child is doing after seeing hundreds of pictures over the course of the past year on Facebook or Instagram.

Constant sharing of information is changing the way our society views interactions, but also our assumptions of what is really happening in society. It is rapidly becoming more acceptable to post frequent updates about oneself on sites such as Facebook, Twitter, and Tumblr. Prior to such online outlets, it was rare to hear such frequent and personal updates from friends. Personal details are now shared repeatedly and often contain an underlying hope of encouragement or approval from one's friends/followers. It has become very popular to "add" as many friends, coworkers, family members, and past acquaintances as friends on these sites as possible to gain online popularity. These have some differences in the various websites. For example, these contacts are labeled as "followers" in Twitter rather than the more egalitarian label of "friends" in Facebook. The process by which users gain followers is different than that of Facebook and does not require the users to send or accept friend requests nor does it require that users become followers of those following them. Thus, Facebook relationships are reciprocal, while Twitter relationships are not (Davenport et al., 2014). Regardless, the desire to be followed or friended online has increased across the population, because it has become the cultural norm to do so.

Some conclude that we are not living in a narcissistic world, but that this is an era of boosting self-esteem. Self-esteem differs from narcissism in that it represents an attitude built on accomplishments we've mastered, values we've adhered to, and care we've shown toward others. Narcissism, conversely, is often based on a fear of failure or weakness, a focus on one's self, an unhealthy drive to be seen as the best, and a deep-seated insecurity and underlying feeling of inadequacy (Firestone, 2012). With that statement, one must wonder why people in today's world are sharing their information. Is it because they are proud of their various accomplishments? It is after all, exciting to share that you got a promotion at your job complete with a pay increase and company stock options. But, does this need to be shared with your entire internet following, and if so…why? A social constructionist approach assumes that all knowledge is created in community through discourse. Simply put, we develop our self-image and our view of others through the particular context of our relationships (Coady et al., 2008, p.376). These types of posts can ignite jealousy in our friends/followers, but also create a sense of superiority and belief that they are jealous, a key factor in diagnosing NPD.

Frequent sharing of pictures online can lead to the thoughts of self-grandiosity from constant feedback from peers. The term "selfie" has become quite popular in the past year, which is a picture that one takes of oneself and then posts online. Peers have the opportunity to "like" or comment on the photos. There's even an option to share the photograph over a multitude of social media websites, increasing the chance of obtaining likes and compliments. This constant pattern of receiving positive feedback can become addicting and some users post endlessly, looking for more feedback to receive the approval they desire. These comments start to create a false sense of superiority, i.e. "I'm so pretty and fabulous, and everyone loves me." While many may in fact think this person is lovely, they may be providing the feedback only because everyone else is doing so. Therefore, following modern society's cues to approve one's picture. The user, however, only sees this as fulfillment to their need of excessive admiration.

Americans may be more narcissistic now than ever, but narcissism is not evenly distributed across social strata. Several studies demonstrated that higher social class is associated with increased entitlement and narcissism (Piff, 2013, p.34). This growth in narcissism in the higher social class may be in part from the rise in reality television shows. Today there are tens of hundreds of "reality" televisions shows that feature the lives of the rich and fabulous. In these shows money and power are glamorized and are implied as being easily obtainable. From the game shows with an end of a million-dollar prize to the housewives of Beverly Hills, it is easy for one to turn on the television and blindly watch and interpret wealth as success. Those who are not quite as wealthy, but may be for their geographic location, may start to believe they should be seen as important members of society and better than others in their area due to their wealth. Piff states, "beyond levels of absolute wealth, individuals who feel particularly wealthy relative to others may exhibit the greatest levels of entitlement and narcissism" (2013, p. 40). This belief can spread throughout groups of people and change the views of how individuals see themselves within their culture. Unrealistic expectations arise and all of a sudden the upper-middle class has a view that they should be open to benefits that only the high upper class members may receive. These unrealistic expectations are also characteristic of NPD.

These shows not only glamorize money, wealth, and class. They often portray a haughty and arrogant attitude. There is an overconfidence depicted, which features these "stars" in televisions series believing they are better than everyone else and above general societal expectations, i.e. DUIs without trial, etc. Frequent viewings of these shows can imbed these same thoughts into the minds of the viewer, who now believes they can get away with saying and doing things that are generally not socially acceptable.

Along with this thought, comes the ability to empathize with others. Empathy is the intellectual identifiction with or vicarious experiencing of the feelings, thoughts, or attitud es of another. These shows often depict men and women who are feuding and finding drama in unnecessary situations. Watching these reality television shows may give the viewer a false sense of supremacy within these people. They may believe it is not necessary to identify with others, but instead, to assume an all-knowing attitude and desire to feud with others instead of trying to come to an understanding of one another. Often these characters on the shows argue in person and online, leaving hostile, aggressive comments on Facebook and Twitter for others to read. Seeing this trend on television, gives a false sense of how society interacts, and viewers start to handle their conflicts in the same manner.

Conclusion

Narcissistic personality disorder is peculiar and leaves one curious as to if it should really be considered a mental disorder. As long as society is promoting self-inflations in television shows and encouraging constant, shameless self-promotion online, this lifestyle will continue to exist. All knowledge is created through community through dialogue. These various exchanges within society often lead to the various criteria for NPD, leaving us to wonder if NPD is a disorder, or just a state of mind. Until further research is done and we can see the implications on society as social media and reality televi sion die down (if ever), no scientific conclusions can be made. However, based on a social constructivist lens, it is my belief that narcissism is learned through society, and not a disorder.

References

American Psychiatric Association., & American Psychiatric Association. (2013).Diagnostic

and statistical manual of mental disorders: DSM-5. Washington, D.C: American Psychiatric Association.

Coady, N., & Lehmann, P. (Eds.). (2008). Theoretical perspectives for direct social work practice. 2nd ed. New York, NY: Springer.

Davenport SW, Bergman SM, Bergman JZ, Fearrington ME. Twitter versus Facebook: Exploring the role of narcissism in the motives and usage of different social media platforms. Computers in Human Behavior. 2014.

Firestone, L. (2012, Nov 16). Is Social Media to Blame for the Rise in Narcissism? Retrieved from http://http://www.psychalive.org/

Kohut, H. (2009). The analysis of the self: A systematic approach to the psychoanalytic treatment of narcissistic personality disorders. Chicago: The University of Chicago Press.

Piff, P. K. (2014). Wealth and the Inflated Self Class, Entitlement, and Narcissism. Personality and Social Psychology Bulletin, 40(1), 34-43.

Ronningstam, E. (2005). Identifying and understanding the narcissistic personality. New York: Oxford University Press.

Sutherland, O., Fine, M., & Ashbourne, L. (July 01, 2013). Core Competencies in Social Constructionist Supervision?. Journal of Marital and Family Therapy, 39, 3, 373-387.

Paraphilic Disorders in Women: Mental Illness or a Reaction to Untreated Trauma?

Lauren Davis

As stated by Gabbard (2005), there are very few psychiatric conditions that carry more negative stigma than paraphilias (p. 313). A disorder that implies a lack of morals and a serious deviation from the norm, the criteria have altered greatly over the years as different sexual practices and preferences have gained normalcy. Currently, paraphilias include sexual behaviors that consist of using nonhuman objects or in which humiliation, pain, children or non-consenting adults are used (Gabbard, 314). Often times, men receive the most attention for paraphilic disorders, but in this chapter I will take a look at paraphilic disorders in women through a feminist perspective. I will argue that labeling a woman with a history of sexual abuse or trauma as a paraphilic is problematic because it does not address the distinction between a psychiatric disorder and social victimization.

The DSM-5 defines a paraphilia as "an intense and persistent sexual interest other than a sexual interest in genital stimulation or preparatory fondling with phenotypically normal, physically mature, consenting human partners," (685). The paraphilic disorders explained in the DSM-5 include voyeuristic disorder, exhibitionistic disorder, frotteuristic disorder, sexual masochism, sexual sadism, pedophilic disorder, fetishistic disorder, and transvestic disorder. Voyeuristic disorder consists of experiencing sexual arousal from watching other people have sex, watching people who are naked and unaware that they are being watched or people getting undressed. Exhibitionistic Disorder is the contrary in which the diagnosed individual is aroused from showing their own genitals to unsuspecting people. Frotteuristic Disorder takes it a step further in which the individual is sexually aroused by touching non-consenting individuals. Sexual Masochism Disorder and Sexual Sadism Disorder consist of being sexually aroused by acts of suffering and humiliation either to oneself or to others, respectively. Pedophilic Disorder is diagnosed when an adult is sexually aroused by children 13 years or younger and has acted on these urges. Fetishistic Disorder occurs when using nonliving objects during sex or having a highly specific focus on a non-genital body part that sexually arouses one. Lastly, Transvestic Disorder occurs when cross-dressing arouses a person. The two criteria for diagnosing a paraphilic disorder include possessing the qualitative nature of a paraphilia and having the paraphilia result in negative consequences in one's life (American Psychiatric Association, 2013, p. 686).

In critically examining the various paraphilic disorders described in the DSM-5, childhood abuse is cited periodically as a potential risk factor. Using feminist theory I will attempt to break down the patriarchal barrier and see how abuse and trauma that goes untreated can cause women to act out in a way that could be perceived as a paraphilic disorder. In applying the components of feminist theory to paraphilic disorder, particularly in terms of oppression, it would appear that often times, women suffering from a paraphilic disorder are actually survivors of sexual trauma that have gone without proper treatment. Coady and Lehman (2008) explain the complexities of feminism stating that for social change to ensue, women must heal at an individual level within a feminist framework (345). Reading the accounts of women with paraphilic disorders, the individual healing did not occur, which later manifested itself as a paraphilic disorder. Countless female paraphilics admit to being victimized by the men in their lives and have been denied adequate recognition when they sought to address the abuse. Patriarchal power appeared to dominate their social justice.

Hypersexuality is not included in the DSM-5 but during the DSM-IV revision, it was proposed and given a distinct list of criteria (Kaplan and Krueger, 2010). The Hypersexual Behavior Inventory (HBI) aims to assess how severely individuals experience emotional discomfort, an inability to control sexual thoughts and behaviors and the degree to which these habits impact their life. Within the proposed diagnosis there are several subtypes including masturbation, pornography, cybersex, telephone sex, and strip clubs. Very limited research has been done regarding hypersexuality in women. Not surprisingly, however, it has been found that at the root of compulsive sexual behavior is "an attempt to recover from early negative childhood experiences" (Kaplan and Krueger, 2010).

As an instructor and therapist at an after school youth program in the South Side of Chicago, I treated a 14-year-old girl, Megan*, [1]who had been severely sexually traumatized by her brother when she was eight-years-old. He was 17 at the time. The program itself is the first known sex education program created by African American youth for African American youth. Seventeen high school students met four days a week to delve into the complexities that face the sex lives of teenagers growing up in an impoverished, violent community where teen pregnancy, sexually transmitted diseases and dating violence are exponentially more prevalent. While a majority of the time in the after school program was spent constructing a curriculum to be taught to other groups of students as well as group discussions and therapy, some time for individual sessions was allotted.

Megan was someone I worked with extensively in one-on-one sessions. In our first session, I asked her to write me a letter about her life experiences, things that made her happy, things that made her sad, and her aspirations for the future. In her letter she explained the traumatic incident that had occurred six years ago. Megan had never re-

---

[1] *Name has been changed for confidentiality purposes.

ceived any counseling or treatment for the incident and, in fact, when she told her parents, they traumatized her further by both doubting and blaming her.

Prior to even learning about the incident, I noticed the lingering effects of child sexual abuse in Megan's overt and inappropriate sexual behavior that she exemplified in social settings. Kaplan and Krueger (2010) cite a few of the symptoms of Hypersexual Disorder as repetitively engaging in sexual behaviors while disregarding the risk for physical or emotional harm to self or others, repetitively engaging in sexual fantasies, urges or behaviors in response to stressful life events, and time consumed by sexual fantasies, urges or behaviors repetitively interfering with other important (non-sexual) goals, activities, and obligations (182).

Megan's hypersexual behaviors were demonstrated during group discussions when her curiosity about specific sexual acts was heightened to a point of concern for instructors. Additionally, she was often seen acting out sexual actions or movements specifically in the presence of men. This behavior was observable over the course of an eight-month period. While it is impossible to know whether or not Megan will develop what would be deemed a paraphilic disorder in the future, she presents one of the biggest risk factors; childhood sexual abuse. Diagnosing her with hypersexuality now or any sort of paraphilic disorder later would be simply placing a diagnosis on an otherwise untreated sexual trauma.

One paraphilic disorder known as pedophilia was the focus of a case study conducted by Chow and Choy on a 23-year-old female, Ms. A., who had strong sexual urges towards young female girls ages 3-4. On two separate incidences she admitted to performing oral sex on girls of this age. Carrying around excessive shame and guilt about her actions in addition to harming these young girls, Ms. A fit the criteria for pedophilia which includes having fantasies or urges about prepubescent children, acting on these fantasies or urges in addition to having them cause personal distress (Chow and Choy, 2002, 214). The person suffering from the disorder must be 16-years of age or older and be at least 5 years older than the victim (DSM-5, 2013, 697). Ms. A. fit these criteria.

The study delved into the circumstances of Ms. A's life and discovered that she had in fact been a victim of sexual abuse and harassment by her stepfather who repeatedly tried to kiss her and fondle her breasts (Chow and Choy, 2002, 213). The allegations went unaddressed by her mother, which created strain and unrest within their relationship. Moving forward, Ms. A suffered emotionally, academically, and interpersonally. Becoming pregnant at age 15, her young adult life was crowded with struggles and challenging circumstances. Still, her childhood abuse went untreated and by 23 she met the criteria for pedophilia.

Following her initial assessment she was prescribed 50 mg of a trial SSRI. While minimal counseling was provided to help Ms. A manage certain stressful situations in her life, no cognitive behavioral therapy was conducted as it was claimed that there are no CBT programs for female sex offenders which again points to the patriarchal advantage in this realm of the mental health system (Chow and Choy, 2002). This is just one example pointing to an unfair treatment procedure in which therapeutic intervention did not occur due to a lack of services for females.

In 1999, a rare case series was conducted in which 14 women with presumed paraphilic disorders were critically examined. Fedorff (1999) states that there is a bias against the study of paraphilias in women both in terms of research and diagnosis. There were 14 women who participated in the study all of whom had been accused of or admitted to exhibiting behaviors indicative of a paraphilic disorder. Within the 14, eight had a specific paraphilia; four could possibly be consider paraphilic and two could not (Fedorff, 1999, 129). Of the sample, the most prevalent diagnoses were pedophilia, sexual sadism, and exhibitionism. Numerous other disorders were found among the participants including anxiety disorders and post-traumatic stress disorder. Fedorff states that with juvenile sex offenders, a history of childhood sexual abuse is often reported. One can also conclude that some trauma has occurred in individuals experiencing anxiety or PTSD.

Another young participant who had been arrested for attempting to tie-up and rape a man claimed during her interview that she believed her mother was trying to poison her when she was growing up. Another woman who was diagnosed as a pedophile was found guilty of two cases of sexual assault in which she committed horrible sexual acts against a 4-year old girl and a 9-year old boy. When asked why she did it, she responded that she was experiencing severe depression and was upset that her mother had neglected her. She also admitted that her mother's boyfriend would force her to watch him engage in sexual activities with her mother. She stated that this had been arousing to her and in her own relationships she was frequently attracted to men who had also been accused of sexual assault.

The fifth case evaluated was another woman who had sexually abused both her sons at ages 1 and 3. In addition to fitting the criteria for paraphilia she also met the criteria for DSM-IV Major Depression. When her depression was treated, her pedophilic behaviors discontinued (Federoff, 1999). The study goes on to cite pedophiles and exhibitionists who had been sexually abused as children.

While exploring these examples through a feminist lens, it becomes apparent that women's mental health needs are being neglected at the price of a diagnosis. Cryle and Downing (2009) explain how women in the 18th and 19th centuries were labeled as criminals for demonstrating overt or seemingly abnormal sexual behaviors. Spencer and Wood (2001) also explain how feminists have strived to disregard the effect of the problematic stereotypes and gender roles history has created. In the Philadelphia Magdalen Asylum "fallen" women who had engaged in sexual activities out of wedlock were sent to live to be governed by a board of men (Spencer and Wood, 2001). At the same time, men who had sex out of wedlock were not stigmatized or otherwise given the label of "fallen". Moving forward men were identified as "public, active, dominant, rational, and cultural where women were identified as domestic, passive, subordinate, and emotional," (Wood, 97, 2001). Placing these ideologies on women creates conflict with the idea of women having disordered sexual behavior. By labeling them with a diagnosis, men are able to displace the responsibility of a patriarchal society in which women are sexually traumatized overwhelmingly more than men. Instead of looking at women diagnosed with paraphilia as women suffering from the effects of sexual trauma, stigmas are reinforced about women with paraphilic disorders, and the root cause goes unacknowledged. Nicki (2001) explains that the norms for mental health for men and women differ drastically. When women demonstrate aggressive behaviors they are often times labeled as mentally ill. The sexist norms eliminate the "personhood" focusing strictly on the mental illness label.

Coady and Lehman (2008) explain how socialist feminists strive to end women's oppression specifically gender-specific features of oppression, namely, sexual abuse and distribution of monetary resources. While child abuse occurs in all walks of life, one can infer when money and education are limited, access to mental health intervention may also be limited. This was indefinitely the case for Megan, as she did not have transportation or resources to access counseling services. It is likely based on her circumstances and the oppression that she faces that her sexual trauma will go untreated. The intersection of Megan's gender and race further amplifies the negative circumstances working against her. Socialist feminists have recently begun addressing this.

Looking at feminism through a radical lens brings even more to the forefront of women's paraphilia issues. Coady and Lehmann

(2012) state, "radical feminists argue that individual women's experiences of injustice and the miseries that women think of as personal problems are actually political issues, grounded in power imbalances," (349). Sexism provides men with the opportunity to have power over women systemically, politically, and mentally. Radical feminism explains male supremacy as having two facets. The first one is that women are damaged psychologically simply by internalizing male supremacy and oppression. The second facet states that as part of the patriarchal system, men are in control of women psychologically and in studying violence against women, control becomes even more apparent according to radical feminists. They point out that violence and men almost always cause abuse towards women. Due to this power imbalance, feminists have strived to develop new ways of treating women's health and mental health issues. Nicki (2001) argues that many psychiatric disabilities are closely linked to traumatic experiences (80). She bases her argument off of Freud's discovery that women who experience hysteria and mental illness were often victims of early sexual trauma (Nicki, 2001, 80).

In exploring various facets of feminism and applying them to paraphilic disorders in women, it becomes apparent that practitioners cannot stop at the diagnosis. The presence of comorbidity in people with paraphilias experiencing depression, anxiety and PTSD point to the indisputable idea that the issue goes far beyond the paraphilia. Research, case studies, and my own clinical experience have demonstrated the negligence and disservice this diagnosis is doing to its recipients. If a woman was abused or traumatized sexually or physically as a child and it goes untreated or even unrecognized they cannot be diagnosed as paraphilic.

Paraphilic behavior and hypersexual behavior are responses to untreated trauma, not a mental illness. This is not to say that the behaviors are not destructive and deserving of serious psychiatric intervention but the paraphilia cannot be treated if the underlying issues go unaddressed. Nicki (2001) addresses the frequent idea that women diagnosed with mental illness are labeled crazy, irrational or stupid. The aim of calling someone any of these names is to eliminate their voice, mute them so that society does not have to deal with the underlying issues. In a typical paraphilic's case, this is abuse. However, this idea of "crazy making" goes for any mental illness, not just paraphilia and confirms the need for feminist centered goals in the mental health world. Without advocating for the proper treatment of women's mental health issues, those suffering will be continuously stigmatized on the bases of the unfounded ideals of what it means to be feminine. In Fedoroff's study, many participants were given psychotropic drugs to help manage their symptoms. While this served as a temporary fix, it is not the solution. Psychodynamic therapy must be conducted in many if not all of paraphilic cases because paraphilia is often times a symptom of a much greater underlying problem.

Nicki (2001) states that "in the case of a person suffering from a mental illness, the claim that her illness is 'all in the mind' would be in one sense correct. Her condition is fundamentally constituted in her mind, in negative thoughts about herself, about her worth and value, about her life and future, possible about others and their lives, or about the world in general as hopelessly evil," (95). A child who was sexually traumatized by the people that were supposed to love and protect her will without doubt feel a loss of control and helplessness. In reading about the ways in which women with paraphilia acted out their urges, it is evident that a majority of them are making ruthless attempts feel in control of others. Whether this be tying up a partner and forcing him to have sex, molesting a young child or seeking reactions from strangers in response to exposing body parts, all of these actions are attempts to feel a sense of power and authority in ways power and authority was had over them.

When you look at a paraphilic disorder from a feminist perspective, oppression, stigmas and inequality must all be considered. In a power struggle to achieve some sense of control particularly for women who have felt that control was taken from them, paraphilia could be a dramatic and devastating reaction to untreated trauma that is now disguised as a mental illness.

References

American Psychiatric Association. (2013). *Diagnostic and statistical manual of mental disorders* (5th ed.). Washington, DC: American Psychiatric Association.

Coady, N., & Lehmann, P. (Eds.). (2008). *Theoretical perspectives for direct social work practice*. 2nd ed. New York, NY: Springer.

Gabbard, G. (2005). *Psychodynamic psychiatry in clinical practice* (4th ed). Washington, D.C.: American Psychiatric Press.

Nicki, Andrea. (2001) *"The Abused Mind: Feminist Theory, Psychiatric Disability, and Trauma."* Hypatia: A Journal of Feminist Philosophy 16.4 80-104. Print.

Fedoroff, J. Paul., Fishell, Alicja, & Fedoroff, Beverly (2001)

"A Case Series of Women Evaluated for Paraphilic Sexual Disorders" The Canadian Journal of Human Sexuality, 8.2, 127-140

Meg S. Kaplan and Richard B. Krueger (2010) *"Diagnosis, Assessment, and Treatment of Hypersexuality"* The Journal of Sex Research, Vol. 47, No. 2/3, Annual Review of Sex Research pp. 181-198

Suzanne M. Spencer-Wood (2001) What Difference Does Feminist Theory Make? *International Journal of Historical Archaeology*, Vol. 5, No. 1 pp. 97-114

Chow, Eva, Chow, Alberto. (2002) "Clinical Characteristics and Treatment Response to SSRI in a Female Pedophile" *Archives of Sexual Behavior* 31.2 211-215

Cryle, Peter, Downing, Lisa. (2009) *"Female Sexual Pathologies"* Feminine Sexual Pathologies 1-7.

Bipolar Disorder and
Social Constructionism Theory

Agatha Demarchi

"Be the change that you want to see in the world."

--Mahatma Gandhi

About 2.4% of people around the world have had a diagnostic of bipolar disorder at some point of their lifetime (www. nimh.nih.gov/statistics). The United States of America has the highest lifetime rate of bipolar disorder at 4.4% (www. nimh.nih.gov/statistics). This paper will explain the definition of Bipolar Disorder by the Diagnostic and Statistical Manual of Mental Disorders (DSM-5), and how some researchers are connecting this disorder with other relevant topics such obesity, early indicators, abuse of medicine and services needed for parents with bipolar disorders.

Thus, this writer will discuss some findings about Bipolar Disorder and how the Social Construction Theory is relevant to the diagnosis with a focus on the limitations of the DMS-5 presents in relationship with this important diagnosis.

Bipolar Disorder Definition

According to the National Institute of Mental Health (NIMH), "Bipolar disorder is a severe form of mental illness with a primary disruption in mood" (www. nimh.nih.gov/statistics). The definition that this document will use for Bipolar Disorder is that explained in the DSM-5, p.123, "bipolar disorder is a brain disorder that causes unusual shifts in mood, energy, activity levels, and the ability to carry out daily tasks".

This illness can affect different aspects in everyday life such as: intimate relationships, poor job or school performance, and can even increase the risk of suicide. However, people with bipolar disorder can be treated and they can have productive lives (www. nimh.nih.gov/statistics).

There are four types of bipolar disorder: Bipolar I, Bipolar II, Bipolar not otherwise specified (BP-NOS), and Cyclothymic Disorder (DSM-5, 2013, p.123).

Bipolar I Disorder is defined by manic or mixed episodes that last at least seven days (DSM-5, 2013, p. 127). "Persistently elevated, expansive, or irritable mood and persistently increased activity or energy that is present for most of the day, nearly every day, for a period of a least 1 week" (DSM-5, 2013, p. 127). Bipolar II, is defined by a pattern of depressive episodes and hypomanic episodes, the difference between Bipolar I and Bipolar II is the presence of maniac episodes (DSM-5, 2013, p.123-138).

A maniac episode is "a distinct period of abnormally and persistently elevated, expansive, or irritable mood and abnormally and persistently increased goal-directed activity or energy, lasting at least one week and present most of the day, nearly every day (or any duration if hospitalization is necessary)" (DSM-5, 2013, p. 124). Moreover, the difference between a manic episode and a hypomanic episode is the presence of functional impairment (DSM-5, 2013, p. 124). Bipolar Disorder not Otherwise Specified (BP-NOS) is diagnosed when symptoms of the illness exist but do not meet diagnostic criteria for either Bipolar I or Bipolar II. However, the symptoms are clearly out of the person's normal range of behavior. (http://www.nimh.nih.gov/health/publications/bipolar-disorder-in-adults/index.shtml).

Lastly, Cyclothymic disorder is a mild form of bipolar disorder (DSM-5, 2013, p. 123). This diagnosis is given to people with both hypomanic and depressive periods for least 2 years (DSM-5, 2013, p.123). "However, the symptoms do not meet the diagnostic requirements for any other type of bipolar disorder." (http://www.nimh.nih.gov/health/publications/bipolar-disorder-in-adults/index.shtml, retrieved on June, 2014).

In addition to symptoms described in the DSM-5, it is important to take into consideration the early indicators for this disorder. Early markers for Bipolar Disorder can include proximal and distal indicators (Benti, Manicavasagar, Proudfoot, Parker, 2014). As these authors explain, "the proximal indicators can include mood fluctuations, sleep disturbance, irritability, and anger, anxiety, racing thoughts and increase of energy" (Benti, Manicavasagar, Proudfoot, Parker, 2014). But these indicators have low specificity and are shared by other psychiatric disorders (Benti, Manicavasagar, Proudfoot, Parker, 2014).

Distal Markers of this disorder can be more difficult to identify and combine social and psychological factors such as family history of mood disorders, high levels of anxiety, sleep disorders, history of early trauma and/or low care and high overprotective parenting, poor early attachments and characteristics (Benti, Manicavasagar, Proudfoot, Parker, 2014).

Social Constructionist Approach

According to Kendall, 2007, p.16 cited in Zastrow & Kirst-Ashman, 2010, Social Constructionism is "the process by which people's perception of reality is shaped largely by the subjective meaning that they give to an experience from this perspective, little shared reality exists beyond that which people socially create. It is, however, this social constructionism of reality that influences people's beliefs and actions". In other words, how people think or believe related to specific situations as they interact with others becomes what it is real for them. Therefore, it is easy to view the word around us as a physical fact (Zastrow & Kirst-Ashman, 2010).

However, Social Constructionism reveals that "we also apply subjective meanings to our existence and experience. In other words, our experiences don't just happen to us. Good, bad, positive, or negative- we also attach meanings to our reality" (Leon-Guerrero, 2005, p.6, as cited in Zastrow & Kirst-Ashman, 2010).

A positive aspect of the social constructionist approach is that it incorporates the concept of human diversity, a major focus in social work (Zastrow & Kirst-Ashman, 2010, p.371). In the point of view of Zastrow & Kirst-Ashman, 2010, "people learn how they are expected to behave through their interactions with others around them. People's behaviors will differ depending on the vast range of circumstances in which they find themselves. Therefore, human diversity should be accepted and appreciated".

Bipolar Disorder and Social Constructionism Theory

As explained above, the Social Construction Theory is the point of view by which each person sees his or her world; therefore, the interaction with others will be based upon this experience (Zastrow & Kirst-Ashman, 2010). As a personal experience, this writer's point of view about Bipolar Disorder was limited because she was not exposed to, has not interacted with, is not informed about and does not know people with this diagnosis. Thus, the social construction that this writer had about people with this illness was based on myths and lack of information as well. This writer's initial point of view about this complex illness was that nothing can be done to manage bipolar disorder, this diagnosis is easy to diagnose, people who have bipolar disorder spend their lives in psychiatric hospitals, they cannot hold down a job, once bipolar disorder is controlled, people can stop taking their medicine and lastly bipolar symptoms are always triggered by some events. However, those are myths.

This writer had an experience in her job place that helped her understand her bias and her need to expand her worldview as well. "Worldview" is the point of view and the perception that people have of the world; therefore, their behaviors and beliefs are correct according to their "worldview" (Collins, Jordan, Coleman, 2013, p.172). This writer holds a Case Manager position in a not-for-profit organization located in Little Village, Chicago IL. One of her responsibilities is to assess new program participants and establish a strong relationship with participants who enter the program.

The *"Carreras en Salud"* (Careers in Health) program is a bridge program that helps students with remedial classes so that they can be accepted into the City College of Chicago into the nursing program to become a Certified Nursing Assistant, Licensed Practical Nurse or Registered Nurse. This program offers case management support, academic adviser guidance, and employment specialist resources. In order to be part of this program participants need to meet the eligibility requirements which include: age of 18 years old and older, bilingual (English and Spanish), and low income (follow the Workforce Investment Act guidelines). Participants need to meet the eligibility requirements, complete the intake Carreras en Salud application and also the initial assessment made by the Academic advisor and the Case Manager as well.

During one assessment, one of the questions that this writer asked was if a potential participant has any health condition that may affect her participation in the program. This person said, "I do not have any health condition that affects my desire to become a Registered Nurse, but I would want to share with you that I have been diagnosed with Bipolar Disorder during the past 4 years". At that moment, this worker understood the need to learned more about this person, about this illness and also recognized that the DSM-5 is an important book that many professionals consult every day, but that day, with this specific situation, the DSM-5 book was not going to be enough of a resource. Through this experience, this worker realized that is it so important to think out of the box, and avoid the myths and the social constructions that people have about mental illness.

The experience above is one out of thousands similar stories that take place every day in the US society. People with Bipolar Disorder have suffered the label that the society has created. Therefore, this writer proposes to do away to the diagnosis of Bipolar Disorder, because this is a way of labeling people who are socially nonconforming. People spend so much time finding similarities between one individual and another; they spend so much work creating labels. The DMS V is a great example of how professionals spend their time identifying, categorizing and putting under the same diagnosis people with similar characteristics. However, they are missing the opportunity to embrace their uniqueness. To be diagnosed with Bipolar Disorder is a condition that changes people's lives forever. Furthermore, the myths related to this condition increase the level of misconception. Society starts to create new definitions and people feel the need for new "labels" (diagnoses) creating a cycle without end. The Social Constructionism (people's beliefs) is the key factor about how society perceives Bipolar Disorder and people with this diagnostic. This writer's point of view is that avoiding the categorization avoids the label. The most important thing is to avoid the conception that to be different is not acceptable under the current society's worldview.

The theory is important, the books are relevant, and the researches are the validation for the methods that professionals use; however, there are other important values that people need to take into consideration and work hard to improve in any life setting (professional and personal): listening skills, empathy, self-reflection, recognizing the dignity and respect of human beings.

Thus, the social work profession promotes these values and reinforces the importance of the human being. "The Core Values reflect (service, social justice, dignity and worth of the person, importance of the human relationship) principles that flow from them, must be balanced within the context and complexity of the human experience" (NASW, 1996)

DSM-5 and Bipolar Disorder

One of the major issues about the misconception about Bipolar Disorder that this writer identified was in relation with the increases of categories of mental disorders that the DSM-5 has and therefore, people generalize terms and concepts about mental disorders.

The first edition of the DSM, published in 1952, contained merely 106 categories of mental disorder (Burston, 2010). The second, published in 1968, had 182 (Burston, 2010). The third edition, published in 1980, had 265, and the revised third edition, published in 1987, had 292 (Burston, 2010). The fourth edition, 1993, has almost 400 and many new categories have been added to the DSM-5 (Burston, 2010).

Thus, this increased number of categories causes a new social construction for the people: use and abuse of medication, abuse of health insurance, hospitalization, generalization about the disorders, and misconception about what they can or cannot do, fear and in some cases hopelessness in people with a diagnosed mental illness and their family as well.

Another important factor that this writer would like to emphasize is the importance of the cultural values. Related to Bipolar Disorder and how the social constructionism has a big impact on the worldview, there is research that shows that Complementary and Alternative Medicines (CAM) are preferred instead of using the medications or as a complement to the medical prescriptions (Strejilevich, Sarmiento, Scapula, Gil, Martino, Gil, & Gomez-Restrepo, 2013). Per Barnes and Ernst, 1997 cited in Strejilevich, Sarmiento, Scapula, Gil, Martino, Gil, & Gomez-Restrepo,2013, the "Complementary and Alternatives Medicines (CAM) are a heterogeneous group of practices which include several medical and health care practices and products that are not an integral part of conventional medicine due to insufficient proof of their safety and effectiveness".

This study took placed in Argentina and Colombia. Two hundred participants completed the survey, 100 in each center (Strejilevich, Sarmiento, Scapula, Gil, Martino, Gil, & Gomez-Restrepo, 2013). The inclusion criteria were: people diagnosed with Bipolar type I and II according to the DSM, aged between 18 and 65 years old (Strejilevich, Sarmiento, Scapula, Gil, Martino, Gil, & Gomez-Restrepo, 2013). The results reported more than half are using CAM looking forward to reducing suffering due to the illness and half of them considered it useful or very useful (Strejilevich, Sarmiento, Scapula, Gil, Martino, Gil, & Gomez-Restrepo,2013). The most important finding was that all those people from this study who are using CAM were reluctant to inform their psychiatrist about it.

Other activities, such as yoga, can help people in their mental and physical life. There are findings in the research made by Goldstein, Liu, Schaffer, Sala and Blanco (2013) that stated the association between obesity and Bipolar disorder are stronger because when most people gained weight, they showed signs of Depression. Therefore, CAM activities can help as a complementary treatment. Between 1990 and 2002 the used of CAM among the adults in the Unites States increased from 34% to 62% (Strejilevich, Sarmiento, Scapula, Gil, Martino, Gil, & Gomez-Restrepo, 2013).

Thus, this study is another example of how while present clinicians are using the DSM to find a possible diagnosis or label for this illness, people prefer in their worldview to use alternatives to the medical approach, and this increases the dilemma between the myths and facts about Bipolar Disorder.

Another myth that many people believe is that people with Bipolar disorder cannot complete daily activities, including parenting. However, according to Venkataraman and Ackerson (2008), mothers with bipolar disorders present the same level of commitment those mothers without any mental issue. The study finds that mothers with bipolar disorder prioritize their children's needs before their own, showing them love, and being available for them (Venkataraman, Ackerson, 2008). Moreover, these results emphasized their good positive attitude related to this mental illness and the mothers as an example for their children (Venkataraman, Ackerson, 2008). The mothers for this study are facing the same difficulty with their children as mothers without this condition: children discipline's issues. (Venkataraman, Ackerson, 2008). However, with the appropriate treatment and support system, mothers with bipolar disorder can find a balance structuring their daily activities, their home, and work place in order to create a safe and healthy environment for their families.

"You may never know what results come of your action, but if you do nothing there will be no result." Mahatma Ghandi

Bipolar disorder is a mental illness that impacts and changes the life of the person who is diagnosed according to the DSM under this criterion. However, the worldview of this person and their environment could be a determinate factor about how others can perceive the reality of this condition. Therefore, society must avoid assumptions, and should be aware that every person is unique, has weaknesses and strengths and these are the most important things. Social Constructionism is created by people's beliefs and experiences. Therefore, it is the responsibility of society to avoid myths about Bipolar Disorder and be more sensitive to people with this condition.

References

American Psychiatric Association (2012). Diagnostic and Statistical manual of mental disorders (5th Ed.). Washington, DC: American Psychiatric Publishing.

Benti, L., Manicavasagar, V., Proudfoot, J., & Parker, G. (2014). Identifying Early Indicators in Bipolar Disorder: A Qualitative Study. Psychiatric Quarterly, 85(2), 143-153. Doi: 10.1007/S11126-013-9279-X.

Burston, D. (2010). Paediatric Bipolar Disorder: Myths, Realities and Consequences. History & Philosophy of Psychology, 12(2), 1-10.

Collins, D., Jordan, C. & Coleman, H. (2013). An Introduction to family social work (4th Ed.). Belmont, CA: Brooks/Cole.

Goldstein, B., Liu, S., Schaffer, A., Sala, R., & Blanco, C. (2013). Obesity and the Three-Year Longitudinal Course of Bipolar Disorder. Bipolar Disorder, 15(3), 284-293. Doi:10.111/Bdi.12035

Strejilevich, S.A., Sarmiento, M. J., Scapula, M.M., Gil, L.L., Martino, D.J., Gil, J. F., & Gomez-Restrepo, C.C (2013). Complementary and Alternative Medicines Usage in Bipolar Patients from Argentina and Colombia: Associations with Satisfaction and Adherence to Treatment. Journal of Affective Disorders, 149(1-3), 393-397. Doi: 10.1016/J.Jad.2012.08.029

Venkataraman, M., & Ackerson, B.J (2008). Parenting among Mothers with Bipolar Disorder: Strengths, Challenges, and Service Needs. Journal of Family Social Work, 11(4), 389-408

Zastrow, C. & Kirst-Ashman, K. (2010) Understanding Human Behavior and the Social Environment (8th Ed.). Belmont, CA: Brooks/Cole.

National Association of Social Workers. (1996). Code of Ethics of the National Association of Social Worker. Retrieve June 2014, from https://www.naswdc.org/pubs/code/code.aspwww.nimh.nih.gov/statistics, retrieved June, 2014 from https:// www.nimh.nih.gov/statistics

Major Depression or Existential Crisis? Approaching the Depressive Disorders through an Existentialist Perspective

Debbie DiMartino

The *Diagnostic and Statistical Manual of Mental Disorders* (5th ed.; *DSM-5*; American Psychiatric Association [APA], 2013) has evolved as a useful guide for practitioners to thoroughly and accurately assess a client's psychological complaints, formulate a diagnosis, and construct a tailored treatment plan. Perhaps no diagnostic category in this manual is more widely consulted by clinicians than the Depressive Disorders. Major Depression affects 1 in 5 people before adulthood (Dew et al., 2010) and leads to upsetting consequences including poor school and work performance, damaged relationships and families, and poor overall quality of life. Often underdiagnosed and inadequately treated, more long-term consequences include recurrent depressive episodes, severe psychosocial impairment, substance abuse, and suicide (Dew et al., 2010). The DSM-5 attempts to distinguish between depressive disorders and the normal, expected disappointments and imperfections of life. Like most other diagnostic categories, it does so by necessitating that the symptoms cause clinically significant functional impairment across several areas of functioning (APA, 2013).

Upon review of the diagnostic criteria for Major Depressive Disorder, it may be startling to find that we experience all of them at one time or another. Hopelessness, emptiness, lack of interest or pleasure in most activities, weight changes, sleep disturbances, fatigue, guilt, feelings of worthlessness, impaired concentration, or thoughts of death (APA 2013) – perhaps with the exception of the latter, who has not experienced all of these symptoms at some point in time? The

DSM attempts to underscore the difference between the disorder and life experiences by assigning very specific and seemingly arbitrary time frames and numeric criteria to the disorder. For example, 5 or more of the above symptoms occurring for 14 days will give you the diagnosis of Major Depression (APA, 2013); but if lasting for only 13 days, is it then just attributed to normal life experience? If a client has only four of the symptoms, is he then free of the diagnosis? He may instead get a label of Persistent Depressive Disorder, but if less than two years, then perhaps it is "Unspecified Depressive Disorder" (APA, 2013). In fact, the very presence of the "Unspecified" diagnosis allows for just about anyone to leave a therapist's office with a legitimate, billable diagnosis of a depressive disorder, regardless of duration or circumstances. This begs the suggestion that the symptoms of the Depressive Disorders are, collectively, an anomaly that should not be part of an individual's life experience.

While the DSM does its best to differentiate between these feelings and mind states as normal life experiences, versus symptoms of a psychological illness, we can better understand the distinction and, therefore, identify the tipping point between the two, if we examine the diagnosis from the perspective of a theory that can potentially explain the underlying reasons for these mood states. Stalsett, Gude, Ronnestad, and Monsen (2012) studied individuals with treatment-resistant depression and found that many of them expressed concerns involving spirituality, "loneliness, meaninglessness, death, lack of freedom, fear of freedom, guilt and shame" (p. 579). These concerns are the very core of Existential Theory, a philosophy which has evolved into a psychological model grounded in the notion that psychopathology, particularly anxiety and depression, is due to "the core issue of angst about the meaning of life and death" (Stalsett et al., 2012). Given this idea, the Depressive Disorders can be explained, from an existentialist perspective, as a crisis in one or more of the four "Ultimate Concerns": Meaning, Isolation, Freedom, and Death.

Existentialism as a philosophy evolved as mankind's attempt to answer the great cliché: What is the meaning of life? Rene Descartes' famous 17th century proposition, "I think, therefore I am", was not as finite of a statement as it seemed at the time. Over the next three centuries, a group of thinkers emerged that became unsatisfied with the emphasis on "I think" and the lack of attention to the meaning of "I am" (Vitemb, 2012). Thus an important philosophical trend evolved, brought to light by individuals whose purpose was to examine the meaning and truth behind our being and existence in this world.

Existential thought is formulated into the "givens," or truths, of existence: Death, Isolation, Freedom, and Meaninglessness. These are considered to be the Ultimate Concerns of the human condition (Vitemb, 2012). We are all constantly aware, whether consciously or subconsciously, of our ultimate demise. We acknowledge our freedom, yet struggle with the responsibility of constructing our own life path. We live in a crowded world yet we are inherently alone and isolated from one another by our differences and experiences. And finally, we are forced to try to make sense of it all while being thrown into a world without discernible meaning (Vitemb, 2012).

These harsh aspects of the human condition inevitably create internal conflict which is unavoidable and certain to evoke some anguish. This is described very well by therapist Shane Ann Vitemb in her discussion on the influence of existentialism on group therapy (2012). She paints a conflicting picture of a mankind which desires life, structure, contact, protection, and meaning, "and yet underneath us is…'nothingness,' an abyss" (Vitemb, 2012, p. 111). Because these inescapable truths are so discordant with man's inherent desires, a certain level of psychological distress is an inevitable, normative experience, and therefore cannot always be considered pathological. Many individuals are able to navigate these concerns throughout life without the functional impairment that is necessary to constitute a DSM diagnosis for any disorder. These individuals still can be considered to have a satisfactory sense of psychological well-being despite a certain level of angst that manages to settle quietly into the background. For many individuals, however, these conflicts come roaring to the front of the psyche, creating a turmoil so penetrating that it disrupts their thoughts, their relationships, their desires, their sense of self. For various reasons (which Existential therapy aims to uncover), they are unable to reconcile these conflicts and therefore a decreased psychological well-being results. For many, this manifests as the symptoms that constitute the depressive disorders.

Therefore, an understanding of existential thinking has significant ramifications for mental health and psychological well-being. For example, Ali Eryilmaz (2011) describes Self Determination Theory, which states that individuals have three needs which determine our subjective well-being: competence, relatedness, and autonomy. Eryilmaz explains that when these needs are met, individuals are happier and healthier; when not met, there is an "increased risk of pathology and illness" (2011, p. 562). If one ponders these three components of Self Determination Theory, they can be shown to align closely with existential concerns. Relatedness is the answer to isolation, while autonomy likens to freedom. Even competence can, in a way, be described as finding meaning in one's actions; when we are competent and capable at a task, it gives meaning to our existence (as in a career, talent, hobby, etc). Therefore, individuals who do not satisfy these three needs in their lives are likely to experience poor psychological well-being, likely in the form of depressive symptoms.

Existential indifference is another concept which illustrates the relationship between existentialism and mental well-being. Tatjana Schnell (2010) describes existentially indifferent individuals as those who "show low commitment to all sources of meaning...particularly self-knowledge, spirituality...and generativity (p. 351). For these people, "positive affect and satisfaction with life is considerably lower," with higher rates of anxiety and depression, than those who experience their lives as meaningful (Schnell, 2010, p. 353).

Both concepts of Self-Determination and existential indifference demonstrate ways in which un-reconciled conflict in the Ultimate Concerns leads to depression. A person free of depression faces these concerns regularly but is able to successfully relate with others, carve out their own life path, and find meaning in life through success in their vocation. Those who are not able to do so, or who show disinterest in self-knowledge or meaning-finding, are not likely to seek out the sources of these needs (relationships, meaningful careers, individuality, agency) and are thus apt be overcome by the Ultimate Concerns and succumb to a chronic state of anhedonia, lack of pleasure, and feelings of worthlessness and meaninglessness, all which constitute the criteria for a depressive disorder.

No discussion of existentialism would be complete without a discussion of religion and spirituality, for one might suggest that mankind's search for meaning goes hand in hand with its search for a higher power, or one-ness with the universe. Since we have seen how existential thought correlates with psychological well-being, then, when looking at mental illness from an existentialist framework, spirituality can be seen as an essential component of subjective wellness. Therefore, while notably absent from the DSM, an evaluation of one's spirituality is a necessary part of the evaluation of mood disorders.

It is important to note the distinction between spirituality and religiosity. Although they are not mutually exclusive, spirituality requires no adherence to a specific deity or set of rules, regulations, and responsibilities associated with religion (Koenig, 2009). Other authors make a similar distinction. Spirituality includes the capacity for growth and the development of a value system (Eliason & Smaide, 2010), inner strength, "unfolding mystery…and harmonious interconnectedness" (Tuck, 2012). Adherence to religion does not require or guarantee any of the above. Religion is "rooted in established tradition…with rules about conduct that guide life" within a group of individuals with common beliefs (Koenig, 2009, p. 284).

The above distinction is important because of the confusing nature of the research regarding religion, spirituality, and depression. In evaluating the literature, we can find studies showing that individuals attending religious services are shown to have a decreased risk of developing major depression, and have a faster recovery after diagnosis (Rasic et al., 2011; Koenig, 2009; Stalsett et al., 2012; Bekke-Hansen et al., 2013). However, other studies demonstrate higher depression scores among those affiliated with religious groups and those who attend services regularly (Koenig, 2009; Baetz & Toews, 2009). The reason for this lies in another distinction: Intrinsic versus Extrinsic religiosity. Individuals with an intrinsic religious orientation have a genuine internal motivation for their faith and practice, and find their reason for being in their religious beliefs (Baetz & Toews, 2009). These people are, therefore, also inherently spiritual. Those with an extrinsic orientation are motivated by external rewards, such as social status, distraction, self-justification, and distraction, or merely because they feel it is required by family or other sources of pressure (Stalsett et al., 2012). Studies that make the distinction between the two show intrinsic religiosity to be correlated with lower depression (Koenig, 2009; Stalsett et al., 2012; Stanley et al, 2011), while extrinsic religiosity is correlated with not only higher depression (Koenig, 2009) but also neuroticism, racial prejudice, rigidity, and severe psychopathology (Stalsett et al. 2012).

Therefore, we can equate intrinsic religiosity with spirituality and use the terms interchangeably. Identifying extrinsic religiosity as a separate entity, and painting a clear contrast, serves to show how individuals belonging to a religion without finding personal meaning, truth, or support, are more likely to experience depression and have a longer time to recovery. These entities - meaning, truth, and support - are, in fact, the essential components of existential thinking and correspond to the Ultimate Concerns. Therefore, spirituality is a crucial component of any discussion on existentialism. Turning back to the consideration of depression defined, existentially speaking, as an inability to reconcile with the internal conflicts created by the Ultimate Concerns, it becomes clear that spirituality is necessary to the understanding of the roots of depression from an existentialist perspective, and must be considered when evaluating clients presenting with symptoms of depression.

Now that the connections between Existentialism, psychological well-being, and spirituality have been clarified, attention then turns towards the roots of depression as they lie within the four Ultimate Concerns and the conflicts they create. First, depression can be explained as a crisis of meaning-finding. Steger, Oishi, and Kashdan (2009) identify adolescence as the stage where the process of creating a sense of meaning begins. This is a stage already rife with conflict, as they struggle with creating a sense of identity and negotiating their role within society. Erik Erikson (1963) theorized that failure to do so leads to role confusion, marked by uncertainty and discomfort with one's place in society; essentially, there is a sense of a lack of meaning to one's life. Surely, anyone who recalls their adolescence can identify with this sentiment. However, as the individual progresses through this and subsequent stages, meaning continues to "unfold in conjunction with individual development" (Steger et al., 2009). Thus, with ongoing failures at establishing meaning, one is faced with the negative outcomes of each stage. When confronted with role confusion, isolation (stage 6), stagnation (stage 7), or despair (stage 8), without appropriate intervention, symptoms of depression present themselves. Thus, the inability to successfully navigate the conflict of meaning-finding leads to depressive symptomatology at any stage of life. This is supported by studies that suggest that the more meaning people report in their lives, the greater subjective well-being and less depression they experienced in all life stages (Steger et al., 2009).

The Ultimate Concern of Freedom, as it helps to explain symptoms of depression, is better served when coupled with Meaning-Finding than when standing as a separate entity. In seeking meaning, for example, depressed mood can be explained by our sense of lack of control over the pursuit, and a lack of ability to influence events and conditions that impact our lives. Self-Efficacy is an excellent measure of one's sense of Freedom and consequent risk of depression. Cohen and Cairns (2012) define perceived self-efficacy as "a person's belief in their capacity to perform in certain ways that gives them control over events that effect their lives" (p. 314). When an individual has a sense of low self-efficacy, they suffer negative emotions such as depression and anxiety which, when amplified by feelings of severe helplessness, portend a full blown psychological disorder. It is no surprise then, from an existentialist perspective, that helplessness is one of the specific diagnostic criteria for a diagnosis of major depression (APA, 2013).

As man struggles in his search for meaning, he inevitably looks to others to help find the answers, and ultimately learns to seek, crave, and need companionship. This leads to the next Ultimate Concern: Isolation. We are inherently isolated in our individual, self-contained state, moving through life with only ourselves as a guaranteed permanent companion. In this loneliness there is inevitably pain; however, this pain is considered normative. In fact, if we did not yearn for companionship, we might be considered to have a mental disorder, such as Schizoid personality (APA 2013). The normal, inherent strife experienced from loneliness can progress to depression if one's desire for social connections is unfulfilled, or is perhaps sought in a maladaptive manner. The non-depressed person recognizes their inherent isolation and need for companionship, and this self-awareness allows him to reconcile this conflict by seeking supportive relationships. Vitemb (2012) supports this notion, stating that by being aware of our isolation, "we can be more aware of what we can [and cannot] get" in relationships with others (p.108). She also offers group therapy as a main treatment for depression, noting that the group processing of lonely feelings leads to "transcendent moments when we learn that ...a yearning for connectedness exists" (p. 108). In this way, accepting one's aloneness never eliminates the state of isolation, but offers a sense of "freedom or independence" (Vitemb, 2012).

In this great existential concern of Isolation we again find the importance of spirituality in preventing or mitigating depressive symptoms. One of the clearest examples of man's search for community is religion (particularly the intrinsic form). In the literature on the implications of religion on mental health, the social aspects of religious community such as health behaviors, group connectedness, and social resources have been identified as mechanisms through which spirituality influences mental health (Baetz & Toews, 2009). Religious beliefs (again, the intrinsic form) provide support, both social and divine, to help reduce isolation and loneliness, especially during trying times.

Perhaps the most profound, inescapable truth we face is that of our mortality. We are all living in a state of death-anxiety, usually simmering beneath our consciousness and brought to the surface when faced with illness, death of a loved one, or countless news reports of catastrophe, terrorism, or even freak accidents. Many of us have been shaken by the fatal collision that occurs on our very own route to work only 20 minutes after we passed that same spot. These anxieties are natural and inevitable, or can perhaps progress to an anxiety disorder if one is consumed by them. Depression, however, is very common when we personally come face to face with our own mortality in the form of life-threatening or terminal illness, or even simply old age. Depression in patients suffering Acute Coronary Syndrome (ACS), for example, is studied extensively and found to be a common co-morbidity in this patient population (Bekke-Hansen et al., 2013). A more evident example of people facing existential depression due to the mortality concern is the population of patients found in palliative care and hospice settings. A significant component of palliative treatment in end-of-life care deals with addressing depressive symptoms based in "existential and spiritual issues of fears, questions, hopes and dreams" (Dewi, Peters, & Margono, 2013, p. 104). Here again, we find one Ultimate Concern intertwined with the others. Returning to Erikson's stages of development, the final stage of life presents us with the conflict of ego integrity versus despair (Erikson, 1963). When we face Death, we reflect on our life review and hope to realize that we have found Meaning in our lives, carved our own path (Freedom), and formed meaningful relationships with others (Isolation). Failure to achieve this realization at this stage leads to despair, helplessness, bitterness, and regret – a recipe for depression.

The issue of Death, and particularly the correlation between the Death Concern and Depression, cannot be discussed in an existential framework without of course addressing spirituality. For example, the study of ACS patients reveals that as they face their mortality, amongst other existential concerns, they tend to turn to their faith as a means of coping. Furthermore, Bekke-Hansen et al.'s findings (2013) support the notion that greater intrinsic religiosity, spirituality, and existential well-being are associated with fewer depressive symptoms in cardiac patients at least 6 months after experiencing an ACS. In Dewi et al.'s (2013) analysis of Muslim patients with cancer in hospice or palliative care settings, religiosity was seen as a powerful treatment tool in the process of improving quality of life and combating depressive symptoms while facing imminent death. By providing value and existential beliefs such as self-esteem, mastery, and rules for a meaningful life, religiosity served as a strong foundation for patients to accept their illness, and reflect upon their life with a sense of achievement (Dewi et al., 2013). As for older adults who are not facing illness or imminent death, but demonstrating difficulty navigating the ego integrity vs. despair stage of life, data suggest that adults in this stage benefit more from treatment of depression when religion or spirituality is incorporated into their counseling (Stanley et al., 2011).

After this analysis of these Existential Truths and how they underlie the psychopathology of Depressive Disorders, we can better understand the DSM criteria and identify that juncture in a client's life where the symptoms cause "clinically significant distress" and functional impairment (APA, 2013). Furthermore, utilizing Existential Theory as a framework for treatment, we can identify clients that might benefit from incorporation of spirituality into our interventions. Certainly many other theories and models of psychotherapy provide options for effective treatment, at least in the short term, but no therapist can deny that lying at the root of a client's depression is likely a pervasive existentialist issue that can only be addressed by a discussion of the Ultimate Concerns of life.

References

American Psychiatric Association. (2013). Depressive Disorders. In Diagnostic and statistical manual of mental disorders (5th ed). Arlington, VA: American Psychiatric Publishing.

Baetz, M., & Toews, J. (2009). Clinical implications of research on religion, spirituality, and mental health. *Canadian Journal of Psychiatry, 5*, 292-301.

Bekke-Hansen, S., Pedersen, C.G., Thyguesen, K., Christensen, S., Waelde, L., & Zachariae, R. (2013). The role of religious faith, spirituality and existential considerations among heart patients in a secular society: Relation to depressive symptoms 6 months post acute coronary syndrome. *Journal of Health Psychology, 4*, 1-14.

Cohen, K., & Cairns, D. (2012). Is searching for meaning in life associated with reduced subjective well-being? Confirmation and possible moderators. *Journal of Happiness Studies, 13*, 313-331. DOI 10.1007/s10902-011-9265-7

Dew, R.E., Daniel, S.S., Goldston, D.B., McCall, W.V., Kuchibhatla, M., Schleifer, C., ...Koenig, H.G. (2010). A prospective study of religion/spirituality and depressive symptoms among adolescent psychiatric patients. *Journal of Affective Disorders, 120,* 149-157.

Dewi, T.K., Peters, M.L., & Margono, B.P. (2013). The effect of religiosity mediated by acceptance on quality of life: A study on muslim patients with cancer in palliative care. *Journal of Law and Social Sciences, 2,* 102-109. DOI: 10.5176/2251-2853_2.2.120

Eliason, G.T., Smaide, J.L., Williams, G., & Lepore, M.F. (2010). Existential theory and our search for spirituality. *Journal of Spirituality in Mental Health, 12,* 86-111. DOI: 10.1080/19349631003730068

Erikson, E. (1963). *Childhood and Society.* New York: W.W. & Co., Inc.

Eryilmaz, A. (2011). A model for subjective well-being in adolescence: Need satisfaction and reasons for living. *Social Indicators Research, 107,* 561-574. DOI 10.1007/s11205-011-9863-0

Koenig, H.G. (2009). Research on religion, spirituality, and mental health: A review. *Canadian Journal of Psychiatry, 5,* 283-291.

Rasic, D., Robinson, J.A., Bolton, J., Bienvenu, O.J., & Sareen, J. (2011). Longitudinal relationships of religious worship attendance and spirituality with major depression, anxiety disorders, and suicidal ideation and attempts: Findings from the Baltimore epidemiologic catchment area study. *Journal of Psychiatric Research, 45,* 848-854. DOI:10.1016/j.jpsychires.2010.11.014

Schnell, T. (2010). Existential indifference: Another quality of meaning in life. *Journal of Humanistic Psychology, 50,* 351-373. DOI: 10.1177/0022167809360259

Stalsett, G., Gude, T., Ronnestad, M.H., & Monsen, J.T. (2012). Existential dynamic therapy ("VITA") for treatment-resistant depression with Cluster C disorder: Matched comparison to treatment as usual. *Psychotherapy Research, 22,* 579-591. DOI 10.1080/10503307.2012.692214

Stanley, M.A., Bush, A.L., Camp, M.E., Jameson, J.P., Phillips, L.L., Barber, C.R., ... Cully, J.A. (2011). Older adults' preferences for religion/spirituality in treatment for anxiety and depression. *Aging and Mental Health, 15,* 334-343. DOI: 10.1080/13607863.2010.519326

Steger, M., Oishi, S., & Kashdan, T.B. (2009). Meaning in life across the life span: Levels and correlates of meaning in life from emerging adulthood to older adulthood. *The Journal of Positive Psychology, 4,* 43-52. DOI: 10.1080/17439760802303127

Tuck, I. (2012). A critical review of a spirituality intervention. *Western Journal of Nursing Research, 34,* 712-735. DOI: 10.1177/0193945911433891

Vitemb, S.A. (2012). Existential thinking and its influence on group therapy. *Group, 36,* 107-119.

Oppositional Defiant Disorder

Michele Early

Diagnosis

Oppositional Defiant Disorder (ODD) is a disruptive and impulse-control disorder that has symptoms that are seen in one or more settings. A youth with this disorder displays angry or irritable mood, also symptoms of losing his or her temper fairly often; the slightest thing can annoy this youth. The youth would have a defiant nature that is also argumentative. This youth may argue with authority figures such as teacher or school staff. If a school administrator asks the individual to complete a request, this child may refuse or openly resist that administrator. One sure symptom is this child always has an excuse for why his behavior was defiant; it is never their fault. The youth is also seen as very resentful and vindictive and displays this at least twice within six months. There are different varieties of severity to ODD. If an individual shows symptoms that are confined to one setting, such as only happening in the home or only at school this a mild case of ODD. If symptoms are present in two settings the ODD is seen as moderate. Finally if symptoms are displayed in three or more settings it is seen as severe. Symptoms of ODD may appear as early as preschool years; by the time a child is an adolescent they would have received the diagnosis by now. (DSM-5, 2013).

The DSM-5 (2013) states that the persistence and frequency of these behaviors should be viewed in the context of the person displaying the symptoms. If the person of that particular age displays that type of behavior you then look at the frequency of the symptoms and is it within normal limits. If this disturbance in behavior is having an impact on social, educational, and other areas of functioning the diagnosis needs to be explored. Diagnostic features of ODD that also have to be revealed is that the youth may have been exposed to hostile parenting. Also needs to be observed is the youth interactions with others outside of the home. Sometimes family dynamic are hostile by design and you may not be able to see if it's the family dynamic that is defiant or the child.

Factors to the ODD are temperamental, the child is unable to regulate their emotions. The child may get frustrated very easily. Also environment plays a role in a child diagnosed with ODD, parenting style such as harsh, inconsistent and neglectful may contribute to the child having the defiant disorder. The disorder diagnosis looks the same in different countries and various ethnicity and races (DSM-5, 2013).

Social Learning Theory

"Think of defiance as an opportunity to teach students something new" (Kohn, 1996). If defiance is a way to teach something new that means the disorder is seen as a learned. If you can learn a behavior you can unlearn the behavior. Social Learning theory suggests that behaviors are learned as they interact with their environment. For example, if asked a person in a leadership position how they acquired those skills that person would state that he observed others who they chose to aspire to be like. They would study and watch how they dealt with situations while in a leadership role. You surround yourself which is what you want to become. Therefore an individual seeking to be a leader surround themselves with mentors that would cultivate that behavior. You are a product of your environment with that being said you learn who you are from what you see and experience.

How do different cultural goals and values affect socialization practices? If you put individuals in the jungle will they adapt to that environment or will the environment adapt to the people? For example in the book Lord of the Flies (Golding 1962), do the youth regress to the primitive state of the wild or have the environment adapt to their civilized state. Throughout the book the majority of the children adapt to the uncivilized nature of the uninhabited island. The groups are divided on the issue of nurture versus nature. One set wants to remain true to the values that they have and other group members believe that you have to become savage in order to survive. The dynamics in the book shows how one that in power influences how others within the group behaves. The first power position chose to be civilize and find food from the island and stay true to their value system. Then when the first power position was lost to another child who believe that you should be able to be free and wild. With this new found leadership he leads the group down an animalistic path. The children desert all sense of moral and values that were instilled in them before been deserted. This show that some are more susceptible to their environmental experiences.

Bandura (1971) stated that consequences are automatically and unconsciously strengthened by immediate consequences. For example, if a punishment is given it needs to be swift and certain. With that being said for an impact on the behavior there needs to be quick action to the reaction to the behavior. Social learning theory states that a behavior can be seen but that does not necessarily mean that the individual observing with model that behavior. There are other factors that effective modeling will be demonstrated, and that includes the person being observed is a constant presence in the person's environment.

Studies have shown that some individuals are more susceptible to adverse environment and being in those types of environment they are more prone to be a product of it. Most metropolitan inner city could not show that more of the modeling behavior. Aggression sometimes is a learned behavior and also symptoms of a much bigger issue. Individuals sometimes are in subprime conditions and feels there is no way out of the situation. Hurt and anger are emotions that people use interchangeably often. When speaking to children that are so angry they really are hurt because of various reasons. Children often go unheard and feel like they don't have a voice. Often times the only way they get any attention is when they are aggressive and angry. For example, if a child constantly observes that the child that misbehave gets attention although unfavorable they still are getting a spotlight shined on them (Scott, S. & O'Connor, T 2012) .

Problem Solving

The person in environment perspective helps the client and practitioner to understand their situation past, present and also future. This perspective could be utilize to teach problem solving techniques. Problem solving techniques helps client to understand the problem, if a child is showing aggression towards other students. Does the child understand that there is a problem, if not that is the first step in problem solving. You want the understanding of the problem within the cognitive level of the child. Work with the child to ensure that they have possible solutions to control or become aware of the aggression. Children often feel that they don't have any input in their lives. This sometimes shows up as aggression which is actually frustration. If you give a person a choice in finding a solution to their problem the problem has more of a chance of actually being solved. The person is vested in the positive outcome. Once you have brainstormed on possible solutions to the aggression or controlling the aggression, allow the client the opportunity to choose which solution they would like to set goals for. Once you set goals discuss with client to ensure that goals are attainable. Client carry out the plan of action to the goals, there are milestones to ensure that the plan is indeed working. Also when the client has complete a set of goals, positive reward needs to be given to ensure continuation of plan (Cummins-Wunderlich, K 1988).

Social Learning and Oppositional Defiant Disorder

Oppositional Defiant Disorder (ODD) places blame on parenting and the style of parenting that a child experiences as to why a child may receive this diagnosis. Many children are angry with their living situation and instead of expressing the sorrow or fear they may experience. They show anger ad aggression as an emotion. Children in inner city school settings show behaviors from the Diagnostic Criteria A of the DSM-5 fairly often. The pattern of angry/irritable and argumentative/defiant behavior I have seen a lot of these behaviors in school settings. It is the same set of children that display this sort of oppositional behavior. Although most of the children would not receive the diagnosis of ODD, the question is should they. If a child meet the criteria of four or more of these symptoms for at least 6 months they may be evaluated for the ODD diagnosis. I believe that the culture of their environment enables them to display and exhibit this defiance. Most of these children grow up in single parent household where the parent is stressed because of a number of factors, i.e. unemployment, lack of resources to support their children, and the ignorance of not knowing.

The susceptibility of a child and how they function in different environments guides their emotional regulation (Scott, S. & O'Connor, T 2012). Certain behavior in families such as pro-social often goes unnoticed although noncompliance and aggression are notice and sometimes rewarded. For example, a boy who stands up for himself even in an aggressive manner may be seen as he going to be able to take care of himself, he's not going to let anyone take advantage of him. That aggression can be internalize as a good way to deal with conflict. This child once in a school setting is seen as a behavioral problem. This starts a vicious cycle of dealing with conflict. Their upbringing states that having these aggressive tendency as positive, although in a school setting it's a negative connotation having this aggression. Schools would like students to adapt to the school environment and not vice versa. The conflict with authority then begins for the child. This child first encounter with an environment is their home which reward such actions. Then when the child encounter another environment such as school it's in direct conflict of what they previously were taught. This behavior has become a learned habit.

How children with learned behavior may develop ODD. Children observe their parents upon first entering the world. Parents often display same temperament and behavior after having baby. A child sees a parent that deals with conflict by losing temper quickly that child starts to exhibit that same type of behavior. Another example that is easily displayed is when a parent comes to the school and speaks with the school staff in a disrespectful way that child learns that they too can defy authority. If parents are children first teacher their behavior exhibit will manifest in their children's behavior. Also children observe how a child that is seen as "bad" may get attention. Although the attention may not be seen as positive it is seen as attention to a child who may not be getting any. The child that is angry or aggressive constantly is been taking out of class and getting one on one attention with a school administrator, i.e. security staff, principal. If a child lives in an environment that is neglectful, that type of attention is attention that they may crave because someone is showing an interest.

Before labeling a child with this diagnosis, an assessment of the child's background needs to be evaluated. Are there tools we can use before placing label on child? This child will not have this label of oppositional defiant disorder upon entering adulthood so what is the purpose of it. We need to put everyone in a category instead of dealing with individuals as such an individual. This diagnosis could be used as a crutch for the child. If the child has symptoms of aggression and angry outburst. That child would learn how to manipulate adults when she display those symptoms, "I have ODD". Although she may have ODD is she been held accountable for her actions?

Social learning theory as it applies to Oppositional Defiant Disorder appears to be a diagnosis that is learned. I believe that the labeling of a child with this diagnosis would not help the outcome of the child. If you have an aggressive child you should treat them as you would any angry person. You give them the tools to adapt to conflict. I think parents need to be educated on how to modify the behavior. You utilize the parents to ensure that they help to teach child to learn how to regulate their emotions. Labeling a child with oppositional defiant disorder that child is already at a negative before he even has a chance to show the opposition. You are what people say you are?

References

American Psychiatric Association (2013). *Diagnostic and Statistics Manual of Mental Disorders.* (5th ed). Arlington, VA: American Psychiatric Association.

Bandura, A. (1971) *Social Learning Theory.* General Learning Corporation.

Kohn, A. (1996). *Beyond discipline: From compliance to classroom.* Upper Saddle River, NJ: Prentice Hall

Roth, G., Assor, A., Niemiec, C., Ryan. R., & Deci, E. (2009). The Emotional and Academic Consequences of Parental Conditional Regard: Comparing and Conditional Positive Regard, Conditional Negative Regard, and Autonomy Support as Parenting Practices. *Developmental Psychology, 45*(4), 1119-1142.

Scott, S. & O'Connor, T. (2012). An experimental test of differential susceptibility to parenting among emotional-dysregulated children in a randomized controlled for oppositional behavior. *Journal of Child Psychology & Psychiatry,* 53(1), 1184-1193.

Anorexia Nervosa and the DSM-5:

Including Family Dysfunction as a Risk Factor

Tracy E. Erwin

What is AN?

Anorexia nervosa ("AN") is one type of feeding and eating disorder included in the newly issued DSM-5. It is a complicated, serious illness and has the highest death rate of any psychological disease. (McPhee, 2000). The DSM-5 sets forth 3 basic symptoms that must exist for a diagnosis of AN. These symptoms are: persistent restriction of energy intake leading to a significantly low body weight; intense fear of gaining weight or becoming fat; and a disturbance in self-perceived body weight or shape. (American Psychiatric Association, 2013). The percentage of body mass index (BMI) is used to determine the severity of the disorder. (American Psychiatric Association, 2013).

AN Demographics

The majority of people afflicted with AN are female. In fact, clinical populations reflect a 10:1 female-to-male ratio. (American Psychiatric Association, 2013, p. 341). AN usually begins in adolescence or young adulthood. (American Psychiatric Association, 2013, p. 341). A recent study of 70 women with a history of AN identified the average age of onset as 16.9 years old. (Tozzi, 2002, p. 147).

The teenage years are a time of life characterized by many physical and psychological changes. Physically, puberty begins in adolescence, changing the body from that of a child into a young adult.

(Erikson, 1963). Psychologically, this life stage is marked by the conflict of two opposing personality traits: ego identity v. role confusion. (Erikson, 1963). According to epigenetic theory, teenagers are looking to define who they are, and they are struggling to integrate past and future versions of themselves. (Erikson, 1963). Peer groups also begin to play a major role in the social and emotional life of the teenager. "The growing and developing youths, faced with this physiological revolution within them, and tangible adult tasks ahead of them are now primarily concerned with what they appear to be in the eyes of others…" (Erikson, 1963, p. 261). Moreover, teenagers seek to distance themselves from parental influence and control. This can create anxiety in the parent-child relationship. "In their search for a new sense of continuity and sameness, adolescents have to refight many of the battles of earlier years, even though to do so they must artificially appoint well-meaning people to play the roles of adversaries…" (Erikson, 1963, p. 261).

DSM-5 Risk Factors for AN

Why do some teenagers become anorexic and others do not? Is it the result of individual characteristics, external influences, or both? Experts generally agree that AN does not have only one cause. (McPhee, 2000). Rather, its emergence is the "perfect storm" of varied, intersecting personal and environmental factors that combine to create a vulnerability to the disorder. (McPhee, 2000).

The DSM-5 lists three categories of risk factors for AN. These risk factors are: temperamental; environmental; and genetic and physiological. (American Psychitric Association, 2013, p. 342). Temperamental traits are anxiety disorders or obsessive traits. (American Psychiatric Association, 2013, p. 342). In fact, most anorexics display signs of an anxiety disorder throughout their childhood. (McPhee, 2000).

The DSM-5 also specifies that certain environments, such as "cultures and settings in which thinness is valued," contribute to the development of the disorder. (American Psychiatric Association, 2013, p. 342). These cultures are often affluent and Western. (McPhee, 2000). Thin is the beauty ideal, as reflected in fashion, entertainment and advertising. When viewing this influence through the perspective of the impressionable teenage mind, one must not underestimate the pressures felt by adolescents to conform to this physical standard of beauty. (McPhee, 2000). Unfortunately, the reach of this destructive message has only increased with the rise of the internet and social media. For example, websites exist (often created by teenagers themselves) that provide a forum for anorexic thoughts and behaviors. (Pro Ana & Mia, n.d.). They often include inspirational rules and sayings, such as "Being thin is more important than being healthy" and "Being thin and not eating are signs of true willpower and success." (Pro Ana & Mia, n.d.). Two recent studies report that the use of *pro*-ana websites results in higher body dissatisfaction and worry about weight gain. (Bardone-Cone, et al, 2007; Harper, et al., 2008). Some experts advocate that a macro-level evaluation regarding website content and overall governmental health policies must occur to prevent an increase in AN. (Abbate Daga, et al, 2006).

Finally, the DSM-5 includes genetic and physiological influences as risk factors for developing the disorder. (American Psychiatric Association, 2013). "There is an increased risk of anorexia nervosa and bulimia nervosa among first-degree biological relatives of individuals with the disorder" (American Psychiatric Association, 2013, p. 342). Other biological influences have been cited as increasing the risk of AN. For example, scientists recently discovered that anorexic brains have an increased level of serotonin, which increases anxiety and obsessive behaviors. (McPhee, 2000). Restricting food intake reduces serotonin levels, and therefore, the corresponding feelings of anxiety. (McPhee, 2000). Unfortunately, the brain adjusts to this decrease and creates more serotonin receptors to compensate. This creates a vicious cycle of requiring less and less food in a futile attempt to decrease the feelings of anxiety. (McPhee, 2000).

Applicable Theory

The Person-in-Environment Perspective

From its inception, social work emphasized the necessity of assessing an individual's psychological health in the context of his or her environment. (Cornell, 2006). Social work has always stressed that therapists must include both internal and external factors in their evaluation of a person's difficulties. "The "person-in-environment" perspective acknowledges the need to consider both the struggles within an individual and how they relate to the external, environmental challenges such as poverty, oppression, trauma and violence." (Cornell, 2006, p. 52). This perspective takes a comprehensive view of all of the factors impacting an individual's psychological health.

Family Systems Theory

Dr. Murray Bowen created a theory of human development set in the context of family patterns. Bowen's Family Systems Theory incorporates the person-in-environment perspective and places it in a more intimate setting--the family. In other words, each family member's psychological health is impacted by the bidirectional interactions with other family members. (Bowen Center, n.d.). Dr. Bowen theorizes that this emotional connection arises from the family's need to depend on each other for survival. "The emotional interdependence presumably evolved to promote the cohesiveness and cooperation families require to protect, shelter and feed their members." (Bowen Center, n.d., p.1). Bowen's theory has eight interlocking concepts, which taken together, explain both the positive benefits of healthy family interdependence, and the destructive effects of negative family patterns.

This paper focuses on one of these concepts, differentiation, in the context of diagnosing AN. Bowen states, "Families and other social groups tremendously affect how people think, feel, and act, but individuals vary in their susceptibility to a "group think" and groups vary in the amount of pressure they exert for conformity." (Bowen Center, n.d., p. 4). Therefore, poorly differentiated families have less tolerance for thoughts, feelings and attitudes that differ from the "group think". When viewed with the intersecting identity struggles faced by teenagers, it makes sense that a poorly differentiated family creates a high level of anxiety for the adolescent seeking independence and her own, separate identity. Salvador Minuchin co-wrote a book several years ago analyzing a variety of unhealthy dynamics and their role in the emergence of an anorexic family member. (Minuchin, et al., 1978). The discussion regarding each of these patterns is beyond the scope of this paper. Therefore, this paper discusses only one aspect of family relationships- the level of individual differentiation.

According to Minuchin, "enmeshed" families, or those with a low level of differentiation, frustrate the attempts of individual members in establishing their own separate identity. (Minuchin, et al., 1978). Emotional boundaries are not respected, and family members are overinvolved in each others' personal lives. (Minuchin, et al., 1978). In response, some adolescents (particularly females) attempt to assert control and boundaries in the only arena available to them-- their bodies.

Why include Family Dynamics as an AN Risk Factor in the DSM-5?

For reasons which are not explained, the DSM-5 does not include family dysfunction as a risk factor for AN. It is this author's opinion that it is a mistake to ignore family interaction patterns during the assessment process for AN.

"Risk Factor" v. "Causation"

As part of this discussion, it is initially important to define the term, "risk factor". Due to the complexity of this disorder, and the emotional reactions surrounding the family's role in contributing to it, its vital to understand that a "risk factor" has a much weaker relationship to AN than a "causation". In the context of mental health, a risk factor is defined as, "any behavioral, environmental, or other consideration which increases the likelihood of developing a disease or disorder, or becoming involved in dangerous situations." (Psychology Dictionary, n.d.). In other words, a risk factor only contributes to a person's vulnerability for a condition and makes it more likely to occur. Conversely, "causation" is defined as, "an empirical relationship which exists between two events which can be summed up as one event (the cause) bringing about the occurrence of the other (the effect)." (Psychology Dictionary, n.d.). With causation, the connection between the factor and the disorder is much stronger and more direct.

In the context of AN, mental health experts also acknowledge this important distinction. The Academy of Eating Disorders (AED) recently published a position paper on this very topic which explicitly states, "It is the position of the Academy of Eating Disorders (AED) that whereas family factors can play a role in the genesis and maintenance of eating disorders, current knowledge refutes the idea that they are either the exclusive or even the primary mechanisms that underlie risk". (LeGrange, et al., 2010, p.1). In other words, although family dysfunction is not the sole cause of AN, it is a risk factor contributing to the disorder.

Theory and Research Support

Bowen's Family Systems Theory illustrates the important role that family relationships play in overall mental health. For example, it explains how families with low levels of differentiation experience increased stress and psychological problems. In the case of AN, when the conflicting pressures of an enmeshed family collide with the teenage desire for autonomy and individuation, the anxious teenager may react by using food restriction as a source of power. This theory has been supported in many clinical studies. For example, a recent study concludes that AN sufferers expressly show an above average desire for autonomy. (Brockmeyer, et al., 2012). Frustration of that motive can increase the likelihood of developing AN. (Brockmeyer, et al., 2012).

Another recent study reveals an even stronger connection between negative family patterns and AN. In this study, researchers asked 70 females with a history of AN open-ended questions regarding what they perceived as the causes of their disorder. Each participant's answer was in a narrative form, and they could include as many causes as they thought were relevant. From the patients' perspectives, family dysfunction was the number one cause of their AN. (Tozzi, et al., 2002). The study includes direct quotes from participants such as: "family always very controlling"; "mother dominant, overpowering, negative"; and "a lot of fights and problems". (Tozzi, et al., 2002, p. 147). If patients themselves perceive that family enmeshment directly caused their disorder, shouldn't the DSM at least include this as a risk factor?

Effective Diagnosis is the Goal, not Blame

Over time, there has been shift in thinking about the family's role in the development of AN. Thirty years ago, the family was portrayed as the primary cause of AN. Minuchin's theory of the "anorectic family" blamed parents for their daughters' AN. (Minuchin, et al., 1978). Minuchin theorized that anorectic families displayed specific and predictable traits. (Minuchin, et al., 1978). Based on what is now known about the complexity of this disorder, that characterization is not only unfair, but also inaccurate. (Le Grange, et al., 2010). There is general consensus among mental health experts that AN results from several intersecting factors, not just one cause. (Le Grange, et al., 2010). Eating disorder specialists recently took the position that, "the AED stands firmly against any etiological model of eating disorders in which family influences are seen as the primary cause...and condemns generalizing statements that imply families are to blame for their child's illness." (LeGrange, et al., 2010, p. 1). In addition, some mental health experts question the conclusions and methodologies of research which advocates the family's primary role in causing AN. (Konstantellou, et al., 2012). Recently, some have also argued that distinct dysfunctional patterns do not exist, and it is erroneous to overgeneralize about anorexics and their families. (Spannuth, n.d.). However, even these critics admit that although there are unanswered questions about the specifics of the family role, it is still a contributing risk factor for developing and encouraging anorexic behaviors. (Konstantellou, et al., 2012; Spannuth, n.d.). "...the importance of comprehending the relationship between the eating disorder and the family should not be overlooked." (Spannuth, n.d., p. 5).

Once the issue of blame is set aside, the real question becomes, why is it necessary to include negative family patterns as a risk factor in the diagnostic process? In general, mental health experts uphold the importance of a comprehensive assessment when making a mental health diagnosis. Therefore, to gain a complete understanding of the patient, both internal and external factors must be examined. Although this "person-in-environment" perspective originated in the social work field, other mental health professionals have begun to embrace it in recent years. For example, in a recent publication, psychologists emphasize the necessity of looking at the person in the context of their life in order to understand their difficulties. (Martin, Jr., et al. 2000).

Since therapists acknowledge the impact of the environment on psychological health, isn't a person's family one of the most important components of this external environment? Aren't family relationships one of the primary sources of interpersonal interaction? The DSM-5 seems to contradict itself by including family dysfunction as an environmental risk factor for other mental health disorders. For example, Separation Anxiety Disorder includes the following in its environmental risk factor description, "Parental overprotection and intrusiveness may be associated with separation anxiety disorder" (American Psychiatric Association, 2013, p.193). The DSM acknowledges the family role when diagnosing certain disorders. Why the hesitancy to include it for AN in the face of years of supporting research and clinical studies?

Conclusion

The DSM-5 reflects current understanding that AN is a complex mental health disorder resulting from many intersecting risk factors. The DSM-5 lists three factors that are biological, emotional and environmental in nature. However, the DSM-5 omits another factor which is relevant and necessary for providing a comprehensive diagnosis of AN—dysfunctional family relationships. Based on Bowen Family Systems Theory, the person-in-environment perspective, and clinical research results, its clear that negative family patterns contribute to the onset of AN. By emphasizing that the goal of including the family is not to blame them, but instead to recognize and minimize factors contributing to the disorder, the DSM will become a more accurate tool for diagnosing AN in the future.

References

Abbate Daga, G., Gramaglia, C., Piero, A., & Fassino, S. (2006, June). Eating disorders and the Internet: Cure and Curse. *Eat Weight Disorders*, 2, 68-71. Retrieved from http://www.ncbi.nlm.nih.gov/pubmed/16809973.

American Psychiatric Association. (2013). *Diagnostic and statistical manual of mental disorders* (5th ed.). Arlington, VA: American Psychiatric Publishing, 190-195, 338-345.

Bardone-Cone, A., & Cass, K. (2007). What does viewing pro-anorexia websites do? An experimental examination of websites exposure and moderating effects: a pilot study. *Int'l Journal of Eating Disorders*, Vol. 40, 537-548.

Bowen Center for Study of the Family (n.d.). *Bowen Theory*. Retrieved from http://www.thebowencenter.org//index.html.

Brockmeyer, T., Holtforth, M., Bents, H., Kammerer, A., Herzog, W., & Friederich, H.C. (2013). Interpersonal Motives in Anorexia Nervosa: The Fear of Losing One's Autonomy. *Journal of Clinical Psychology*, Vol. 69(3), 278-289.

Causation. (n.d.). In *Psychology Dictionary*. Retrieved from http://psychologydictionary.org/causation.

Cornell, K. (2006). Person-In-Situation: History, Theory, and New Directions for Social Work Practice. *Praxis*, Vol. 6, 50-56.

Erikson, E. (1963). *Childhood and Society*. New York, NY: W.W. & Co., Inc., Chapter 7: "Eight Ages of Man", 261-263.

Harper, K., Sperry, S., & Thompson, J. (2008). Viewership of pro-eating disorder websites: association with body image and eating disturbances. *Int'l Journal of Eating Disorders*, Vol. 41, 92-95.

Konstantellou, A., Campbell, M. & Eisler, I. (2012). *A Collaberative Approach to Eating Disorders*. New York, NY: Routledge, Chapter 1: "The family context: cause, effect or resource", 5-18.

Le Grange, D., Lock, J., Loeb, K., & Nicholls, D. Academy for eating disorders position paper: The role of the family in eating disorders. *International Journal of Eating Disorders*, Vol. 43(1), 1-5.

Martin, Jr., W. & Swartz-Kulstad, J. (2000). *Person-Environment Psychology and Mental Health: Assessment and Intervention*. Mahwah, NJ: Lawrence Erlbaum Associates.

McPhee, L. (Writer, Producer & Director). (2000). Dying to be Thin [Television Series Episode], *NOVA*. New York, NY: Corporation Public Broadcasting.

Minuchin, S., Rosman, B., & Baker, L. (1978). *Psychosomatic Families: Anorexia Nervosa in Context*, Cambridge, MA: Harvard University Press.

Pro Ana & Mia (n.d.) Retrieved Jul 1, 2013 from http://iloveanamia.webs.com.

Risk Factor. (n.d.). In *Psychology Dictionary*. Retrieved from http://psychologydictionary.org/risk-factor.

Spannuth, W. (n.d.). *Family Structure in Eating Disorders*, Vanderbilt University Psychology Department, Retrieved from http://www.vanderbilt.edu/AnS/psychology/health_psychology/famstruc.htm.

Tozzi, F., Sullivan, P., Fear, J., McKenzie, J. & Bulik, C. (2002). Causes and Recovery in Anorexia Nervosa: The Patient's Perspective. *Wiley Periodicals*, 143-154 doi: 10.1002/eat.10120.

Gender Differences in Borderline Personality Disorder

Andrew Fishman

People with borderline personality disorder (BPD) make up around 15-25% of the people receiving treatment in a given clinical setting (Gabbard, 2005). Many more women are diagnosed with BPD than men; some researchers estimate that women may represent up to three quarters of the population with a diagnosis of BPD (Johnson, et al., 2003). Several studies have been conducted which show gender differences in the diagnosis of BPD (Gabbard, 2005). Some research-ers feel that this wide difference is not present in the general public, that in a random sampling of people in a non-clinical setting, roughly half of the people with BPD would be men and half would be women (Sansone & Sansone, 2011).

A person with BPD can be considered unstable in many aspects of his or her life (American Psychiatric Association, 2013). This can manifest as highly unstable relationships, switching between feeling as though other people are "gods or devils" (Gabbard, 2005). Under-standably, this can result in difficulty maintaining a healthy relation-ship with another person, as any negative interaction has the potential to switch (usually temporarily) the person from "all good" to "all bad" in the mind of a person with BPD.

This instability also results from severe anxiety regarding the re-lationships a person has (Gabbard, 2005). For some with BPD, this anxiety can be such that he or she will threaten or attempt suicide or acts of self-injury in order to ensure the other person will stay in the relationship. Unfortunately, that the other person will suddenly leave the relationship is not the only fear people with BPD tend to have in relationships. Many fear that their relationships will completely con-

sume their own identity. Similarly, people with BPD often have difficulty maintaining a stable sense of who they are, report feeling "empty" and are prone to sudden and inappropriate bursts of anger (American Psychiatric Association, 2013).

Researchers disagree about the reason for discrepancy between
men and women in clinical populations. The most prominent theories
for this difference are:

- Women with BPD may seek treatment more frequently than
  their male counterparts (Sansone & Sansone, 2011, Goodman, et al., 2010).
- Men may be more likely to receive a diagnosis of Antisocial
  Personality Disorder instead of BPD, even when presenting
  with identical symptoms (Gabbard, 2005).
- Men with BPD may be more likely to have other comorbid
  personality disorders, which makes proper diagnosis difficult
  (Johnson, et al., 2003).
- Men may display symptoms of BPD differently; they are especially likely to externalize impulses, while women are
  more likely to internalize these impulses (Johnson, et al.,
  2003).

Although these theories vary greatly, and all may be true, they
each have to do with stereotyped gender roles and expectations in
modern society.

Rates of Treatment-Seeking

Determining whether prevalence of any given mental disorder
differs between men and women is made difficult by our culture's
tendency to encourage women to express their emotions and seek
help, as well as its tendency to discourage the same in men. Goodman, et al., (2010) conducted an anonymous online survey of 495
men and women regarding their use of various treatment facilities in
order to determine whether women with BPD were more likely to uti-

lize such resources than men were. Interestingly, of the 495 surveys collected, more than 400 were completed by women. This stark difference could reflect any number of variables, especially that women may be more likely to answer questions regarding their mental disorders than men. The male participants who responded were significantly less likely to receive psychotherapy or pharmacotherapy over the course of a lifetime than the females were, although they were more likely to utilize drug/alcohol rehabilitation services. The findings suggest that women do seek treatment more frequently than men with similar diagnoses.

This clearly relates to the idea that traditionally, women are expected to be more in tune with their emotions than men are. If a woman were in emotional pain, it would be much more acceptable for her to express these emotions and see a therapist than it would be for a man to do the same. One is reminded of the phrase "man up" as an example of what we expect young boys to do when experiencing pain: suppress it.

Clinician Bias in Diagnoses

It is certainly possible as well that clinicians are biased in their assessments of people with similar or identical symptoms. Several alarming studies have suggested that, even given identical cases, clinicians often diagnose identical cases differently when presented with female names than cases presented with male names (Garb, et al., 1997). These differences are most common in diagnoses which reflect gender stereotypes; histrionic personality disorder was more commonly diagnosed in females, and antisocial personality disorder was more frequently diagnosed in males. Similar research has been conducted with BPD, but results vary between studies, some suggesting a bias and others not.

One such study found significantly higher rates of diagnosed BPD in females than males, even though the presented studies were identical (Becker & Lamb, 1994). Three hundred eleven surveys were collected from psychologists, psychiatrists, and social workers. Each was given a case study to review and was asked how applicable a variety of diagnoses were to the fictional client. Half of the participants received a study with a female name, and the other half received the same study, but with a male name. Significantly more clinicians who believed the patient was female felt the diagnosis of BPD was applicable than did clinicians who believed the patient was male.

This also reflects gender stereotypes, but from the perspective of the clinician. The clinicians may have perceived the aggressive behavior indicated in these studies differently when associated with a male name than when associated with a female name. These subconscious associations also have their roots in a culture which allows men to be aggressive and does not for women. However, these studies do not examine whether men and women with BPD look the same and present with the same symptoms.

Differences in Presentation of Symptoms and Comorbidity

In order to determine whether males and females with BPD may present differently, Johnson, et al. (2003) used the findings of the Collaborative Longitudinal Personality Disorders study conducted by the National Institute of Mental Health as a baseline with which to compare their own surveys. Six hundred sixty eight participants were surveyed on a variety of variables, including the Diagnostic Interview for DSM-IV Personality Disorders, the Structured Clinical Interview for DSM-IV Axis I Disorders – Patient Version, and the Longitudinal Interval Follow-Up Evaluation Base-Line Version. Johnson, et al. found gender differences in many variables for people with BPD, especially Post Traumatic Stress Disorder (PTSD), eating disorders, and substance use disorders. Women were more likely to have PTSD, eating disorders, identity disturbance, and to report a history of physical or sexual abuse. Men were more likely to have substance use disorders and have comorbid diagnoses of various personality disorders. Understandably, these differences can account for differences in accurate diagnoses, especially because the vast majority of research on BPD focuses on women. This could easily account for much of the difference in rates of diagnosis. Because this study suggests that there are definite differences in how BPD presents itself in men and women, and women with BPD are more studied than men with BPD, these women would logically fit the profile of a person with BPD better than their less-studied male counterparts.

A study conducted on 294 adolescents with BPD found similar differences in the presentation of BPD (Bradley, et al., 2005). Like Johnson, et al., Bradley, et al. surveyed the participants using a variety of measures, including the Shedler-Westen Assessment Procedure-200 for Adolescents. Adolescent girls with BPD were more likely to resemble adults with BPD. Adolescent boys with BPD were more likely to be aggressive, disruptive, and antisocial. This is similar to the findings presented in the previous study, which suggest that men with BPD are more likely to have externalizing symptoms than women with BPD.

This, too, has to do with gender roles, in that one of the only acceptable emotions men may express is anger (Parkins, 2012). Women are expected not to externalize anger and so may, consciously or not, work much harder to control their impulses than males would. One need only think of an office with an emotional dispute to get at what this might look like. A man raising his voice to get his point across would likely be seen as "assertive," while a woman doing the same would be seen as "bossy."

Feminist Theory

Each of these possibilities can be accounted for by feminist theory. Several feminist theorists have pointed out that the diagnostic criteria for personality disorders reflect harmful gender stereotypes and would result in different rates of diagnosis between men and women (Serrita Jane, J. & Oltmanns, T.F., 2007). For example, male clients have been rated as being less well-adjusted compared to their female peers when described as "passive," and the opposite was true when clients were described as "aggressive." This clearly mirrors societal expectations of both men and women, in which passive females and aggressive males are the norm.

Additionally, the connection between childhood sexual abuse and BPD cannot be ignored. A 2003 study examining the relationship between early- and late-childhood sexual abuse in women and both PTSD and BPD found, unsurprisingly that 94% of the 38 women who had been abused under the age of thirteen showed signs of both BPD and PTSD (McLean, L.M & Gallop, R., 2003). By stark contrast, 0% of the 27 women who had been abused later in life had developed both BPD and PTSD.

This strong link seems to contradict the above studies that suggest that the numbers of men and women with BPD are roughly even. Because childhood sexual abuse occurs much more often in girls than boys (Briere, J. & Elliott, D.M., 2001), it would logically follow that women would develop BPD more frequently than men would. However, non-sexual physical abuse occurs about as often to young boys as it does to girls. If BPD can be shown to be related to non-sexual physical abuse, it could make up for the difference one would expect to find between genders.

Does Borderline Personality Disorder Exist?

It cannot be denied that there is a large population of people who present with symptoms of BPD. However, whether this means that there is an objective, real diagnosis which we call BPD can be debated. It is also possible that what we call BPD is really a socially constructed way of looking at a common group of symptoms. Further, if these symptoms are different for men and women, can it be said to be a distinct disorder itself? It is certainly possible that this diagnosis is just a way of describing deviation from what is societally accepted.

Might males with BPD be diagnosed less frequently because they blend in more easily into society? The symptoms of BPD related to aggression and impulsive behavior do fit more closely what is expected of males than expected of females. In other words, the same level of aggression in a female might be perceived as pathological, while it could be overlooked in a male. As one Huffington Post contributor wrote, "When a man asserts himself, society calls him a go-getter. People are impressed when a man stands up for himself or achieves his goals because we have been conditioned to believe that it is acceptable for men to be assertive. Women, on the other hand, are supposed to smile and be nice" (Fader, S., 2014). In short, serious work needs to be done to evaluate whether the current diagnostic criteria listed in the DSM are appropriate for both men and women and whether more can be done to educate clinicians about the danger of incorrect diagnosis based on socially constructed gender roles.

References

American Psychiatric Association. (2013). *Diagnostic and statistical manual of mental disorders* (5th ed.). Arlington, VA: American Psychiatric Publishing.

Becker, D. & Lamb, S. (1994). Sex bias in the diagnosis of borderline personality disorder and posttraumatic stress disorder. *Professional Psychology: Research and Practice, 25*(1). doi: 0735-7028/94/S3.00

Briere, J. & Elliott, D.M. (2001). Prevalence and psychological sequelae of self-reported childhood physical and sexual abuse in a general population sample of men and women. *Child Abuse & Neglect, 27.*

Bradley, R., Zittel Conklin, C. & Westen, D. (2005). The borderline personality diagnosis in adolescents: gender differences and subtypes. *Journal of Child Psychology and Psychiatry, 46*(9).

Fader, S. (2014, February 21). Stop calling assertive women b*tches. *Huffington Post.* Retrieved from http://www.huffingtonpost.com/sarah-fader/stop-calling-assertive-women-bitches_b_4808451.html

Gabbard, G.O. (2005). *Psychodynamic psychiatry in clinical practice.* Arlington, VA: American Psychiatric Publishing, Inc.

Goodman, M. et al. (2010). Treatment utilization by gender in patients with borderline personality disorder. *Journal of Psychiatric Practice, 16*(3). doi:10.1097/01.pra.0000375711.47337.27

Johnson, D.M. et al. (2003). Gender differences in borderline personality disorder: Findings from the Collaborative Longitudinal Personality Disorders study. *Comprehensive Psychiatry, 44*(4). doi: 10.1016/S0010-440X(03)0090-7

McLean, L.M. & Gallop, R. (2003). Implications of childhood sexual abuse for adult borderline personality disorder and complex posttraumatic stress disorder. *American Journal of Psychiatry, 160*(2). doi: 10.1176/appi.ajp.160.2.369

Parkins, R. (2012). Gender and emotional expressiveness: An analysis of prosodic features in emotional expression. *Griffith Working Papers in Pragmatics and Intercultural Communication 5*(1).

Sansone & Sansone (2011). Gender patterns in borderline personality disorder. *Innovations in Clinical Neuroscience, 8*(5).

Serrita Jane, J. & Oltmanns, T.F. (2007). Gender bias in diagnostic criteria for personality disorders: An item response theory analysis. *Journal of Abnormal Psychology, 116*(1). doi: 10.1037/0021-843X.116.1.166

A Case Against Remission from Depression: A Rhetorical Critique of the DSM-5

Paige Gesicki

Introduction

If "language is where power is embedded and where the struggle for emancipation can occur," perhaps the right language could emancipate someone from a labeling mental health diagnosis, or even the emotional turmoil itself (Crawford, P., Johnson, A., Brown, B., & Nolan, P., 1999). Often times, words are too removed from lived experience to adequately to describe the intricacies of human emotional phenomena. However, language, a socially constructed system of symbols, is the primary way that humans attempt to communicate their experiences, including a pervasive mood state such as major depressive disorder. Unlike disease states of the physical body, the presence of major depressive disorder lacks biological markers and is therefore impossible to detect through visually observable measures. Thus, the path of major depressive disorder, from its onset through recovery, is often tracked through verbal and written accounts—the development of the illness is relegated to the malleable world of words.

When considering the importance of language as it pertains to mental health, it makes sense to start by examining the Diagnostic and Statistical Manual of Mental Disorders, 5th Edition (DSM-5), the canonical text on mental health diagnoses. Mental health professionals refer to this comprehensive list of diagnostic criteria in the DSM-5 to diagnose individual psychopathologies. Diagnoses are helpful to clinical work because they provide a standardized description of the individual's "problem" which thereafter guides clinical treatment (Ishibashi, 2005).

I will examine how the language of the DSM-5 shapes an individual's experience of his or her mental health state. I use the rhetoric-centric perspective of Narrative Theory to examine the validity of the language used in the diagnostic criteria of major depressive disorder diagnosis in the DSM-5 with particular attention to the term "remission." According to narrative theory, language is a social construct that has the powerful ability to generate meaning in our lives— essentially, the story we hear about ourselves affects the story we tell ourselves, thus creating our reality (Coady & Leahmann, 2008). Narrative theory promotes paying rigorous attention to an individual's word choice and phrasing; I recommend applying these same rhetorical standards to the widely used DSM-5. This paper questions whether the term "remission" should be included in such a prescriptive text as the DSM-5 which, because of its widespread medical authority, has the power to affect a diagnosed individual's sense of self.

I argue that remission is an inappropriate and potentially harmful term to describe a person's relationship to major depressive disorder for numerous reasons. First, measuring recovery on the basis of a decrease of negative symptoms on a standardized list (a score of less than 7 on the Hamilton depression rating scale) is problematic because major depressive disorder cannot be measured objectively, and "in the absence of biological markers of the disease state, remission from depression has been defined phenomenologically" (Zimmerman, McGlinchey, Posternak, Friedman, Attiulah & Boerescu, 2006, p. 148 & Zimmerman, 2010, p. 149). Additionally, "remission" is an idea constructed by a medical and mental health team, yet studies show that people who have actually experienced depression believe the term does not accurately reflect the lived experience of recovery (Zimmerman, et. al., 2006, p. 148). Most importantly, since language has the power to affect one's thought and behaviors, diagnosing someone as "in partial or full remission" risks the possibility of that individual feeling permanently tied to his or her pathological diagnosis.

Studying the intersection of linguistic anthropology and social work, Ishibashi charges clinical social workers with the responsibility of asking, "what influence might the language of diagnosis impose, and how a specific diagnosis, as a label, impacts the way clinicians treat clients" (Ishibashi, 2011, p. 67). By qualifying someone's recovery as "in remission," the person is robbed of his or her ability to be fully healed; the experience of depression is no longer viewed as a transient stage, but instead a fixed state of being or a pathological life sentence.

Narrative theory contends that an individual's problem or condition is socially constructed through language, and therefore the problem can, and should, be resolved through language (Coady & Lehmann, 2008). When applying narrative theory to the experience of clinical depression, the question arises: is it possible to generate a new meaning for the words used in the clinical description of depression, a meaning that is curing and healing instead of incarcerating and heavy? For a narrative theorist, this would entail deconstructing the client's reality and creating a new conversation. As per the theory, it is important to recreate an empowering alternate story *through conversation or discourse* because language gains power in its exchange with another. This paper introduces two examples where narrative theory mitigates against the potentially harmful effects of being diagnosed with major depressive disorder in remission. These two structured treatments—narrative therapy, the clinical practice of narrative theory, and mood memoirs—encourage an individual to be the protagonist of his or her own life story.

Definition of Major Depressive Disorder

In the DSM-5, major depressive disorder "represents the classic condition in this group of (depressive) disorders," unifying the other depressive disorders through the common characteristics of "the presence of sad, empty or irritable mood, accompanied by somatic and cognitive changes that significantly affect the individual's capacity to function" (American Psychological Association, 2013, p. 155). Because it provides the base for all depressive disorders, major depressive disorder is the most illustrative example and will therefore be the focus of this paper. The contrived diagnostic language of the DSM-5 attempts to describe the experience of depression in words that are accessible to mental health professionals through standardization of language, however the words feel too far removed and flat to describe the reality of the psychological phenomenon.

The linguistic shortcomings to describe depression are not limited to the DSM-5. Levitt, Korman & Angus studied the use of "burden" metaphors used by clients in therapy as they attempted to described their experience of depression. The study acknowledges that the subjective experience was so difficult to verbalize that clients turned to descriptive metaphors in their attempt to "more accurately capture of the quality of an emotion (instead of using) an adjective or an emotional label" (Levitt, Korman & Angus, 2000, p. 24).

Additionally, the word "depression" has become so diluted in mainstream English vernacular that it not only grossly fails to describe the lived experience of a clinical depression, but it also has assumed a distinctly different meaning in its common use. Pulitzer Prize winning author William Styron describes the "semantic damage" of the misuse of the word, which "has slithered innocuously through the language like a slug, leaving little trace of its intrinsic malevolence and preventing...a general awareness of the horrible intensity of the disease..." (Styron, 2010). In his memoir, *Visible Darkness,* Styron describes his frustration with the descriptive limits of language, as well as a brief demonstration of the evolution of words in their social construction; he writes:

> "When I was first aware that I had been laid low by the disease (major depressive disorder), I felt a need, among other things, to register a strong protest against the word "depression." Depression, most people know, used to be termed "melancholia," a word which…would still appear to be a far more apt and evocative word for the blacker forms of the disorder, but it was usurped by a noun with a blank tonality and lacking any magisterial presence, used indifferently to describe an economic decline or a rut in the ground, a true wimp of a word for such a major illness" (Styron, 2010).

Styron's memoir is characterized as a mood memoir for its in-depth description of his first hand account of his depression, which he calls "the disease". Styron demonstrates the healing benefits of externalizing the problem, a strategy that is often utilized in narrative therapy. He successfully used his mastery of language to gain enough distance from his debilitating experience of the mood disorder to write an illuminating first-hand account. According to the DSM-5, Styron might have been characterized as in remission from his major depressive disorder, which would be a treatment success, since experts "suggest that achieving remission of symptoms should be viewed as the primary goal (of treatment)" (Zimmerman, Martinez, Attiullah, Friedman, Toba & Boerescu, 2012, p. 78).

The Questionable Basis of "Remission"

If remission is the primary objective of clinical treatment, the question should be posed: what exactly characterizes remission in the realm of mental health? In 1988, the MacArthur Foundation Research Network on the Psychobiology of Depression organized a conference to review and tighten the definition of many "recovery" terms, including the term remission (Frank, Prien, Jarrett, Keller, Kupfer, La-vori, et al., 1991). After acknowledging the considerable inconsistencies in the views of the course of depression, the task force agreed that remission would thereafter refer to a "brief period during which the individual is asymptomatic." Asymptomatic does not mean the complete absence of symptoms. Instead, it is defined as the presence of no more than minimal symptoms, as proven by a score of 7 or lower on the 17-item Hamilton Depression Rating scale (Zimmerman, 2006, p. 148).

Judging a person's recovery on the Hamilton depression rating scale is problematic because the scale has been criticized as an outcome measurement for numerous reasons, including its inability "to define remission with appropriate and empirically derived cutoff points" (Ballesteros, Bobes, Luque, Dal-Re, Ibarra & Guemes, 2007). As stated previously, there are no "empirically driven" biological markers to prove the presence or absence of depression. In fact, the DSM-5 concedes that, "although an extensive literature exists describing neuroanatomical, neuroendocrinological, and neurophysiological correlates of major depressive disorder, no laboratory test has yielded results of sufficient sensitivity and specificity to be used as a diagnostic tool for the disorder" (American Psychological Association, 2013, p. 165).

Across medical fields, if a patient is deemed "in remission" the individual is not necessarily free of the illness or disease at hand; instead, the phrase implies that the illness has abated temporarily, and may return. Therefore, if someone is diagnosed with depression in full remission, the implication is that they have not experienced any "significant signs or symptoms of the disturbance" in the past 2 months (American Psychological Association, 2013, p. 188). The DSM-5 relies even further on the measurability of major depressive disorder by providing the additional qualifier of "partial remission," which is when symptoms are present but full criteria are not met for major depressive disorder (American Psychological Association, 2013, p. 188). Thus, it follows that major depressive disorder, according to the DSM-5, is not a transient mood state from which a person can fully recover. Likewise, the grassroots mental wellness advocacy group NAMI, the National Alliance on Mental Illness, defines depression as "a life-long condition in which periods of wellness alternate with recurrences of illness" (Duckworth & Shelton, 2012).

Although it is true that individuals with major depressive disorder often experience recurring depressive episodes, I contend that adding the qualifier "in remission" to the diagnosis, instead of eliminating the diagnosis entirely, pathologizes a person (Duckworth & Shelton, 2012). The individual is now medically regarded as a "depressive in remission" instead of an individual who once suffered or occasionally suffers from a depressive episode. In terms of narrative theory, "in remission" suggests that the disease will return. This insinuation makes it hard to integrate depression as a past episode of one's life story, due to the looming fear that an episode will return at any given time. Instead of being rid of the pathology altogether, the label remains with a qualifier. In the terms of narrative theory, a client who self-identifies as "in remission" due to the prognosis provided by a clinician provides "life support" and fuel to the problem, therefore keeping the problem alive.

Perhaps an additional problem lies in who has the "last word" in establishing recovery or a lack thereof. In his article entitled "How Should Remission from Depression be Defined? The Depressed Patient's Perspective," Dr. Mark Zimmerman M.D. strays from the traditional "medical model" in favor of a client-centered perspective, which is traditional to the social work paradigm. He found that patients who deemed themselves free of the depressive state reported that the "presence of positive features of mental health such as optimism, vigor, and self-confidence was a better indicator of remission than the absence of the symptoms of depression" (Zimmerman, 2006, pg. 150). Thus, the experts on the depression, the individuals who experienced it themselves, agreed that the DSM's definition does not accurately reflect the lived experience.

Implication for Practice:
Narrative Therapy and Mood Memoirs

The DSM-5's medical model is in line with modernist approaches, which prioritize objectivity and tend to be more diagnostic in nature, often creating a power dynamic with therapist as expert and client as a subject (Ishibashi, 2011, p. 70). Therefore, when a clinician applies a particular diagnosis to an individual, he or she utilizes a rhetorical currency that only privileged professionals can speak with authority, excluding others from the conversation. The resulting power dynamic directly contradicts the National Association of Social Workers Code of Ethics, which states, "Social workers understand that relationships between and among people are an important vehicle for change. Social workers engage people *as partners* in the helping process" (NASW Code of Ethics, 2008).

In line with the egalitarian mission of social work, Narrative theory is *post*modern, and asserts that reality is socially constructed through language, which is comprised of a formal system of symbols (Coady & Lehmann, 2008). Narrative therapy is a clinical model based on narrative theory; it "emphasizes an elaboration of constraining monologues to liberating dialogues and/or the deconstruction or rewriting of problem-saturated stories to stories of courage, strength, and competence" (Coady & Lehmann, 2008). This type of therapy aligns with narrative theory's assertion that language is not an individual endeavor—problems are created through discourse with others and therefore need to be addressed in conversation with others.

This dialogical imperative introduces the second method of narrative healing—the mood memoir. A journal that a writer keeps to him or herself can be cathartic, but explaining an experience in one's own words and sharing that with a willing listener or reader is powerful because of its discourse. Memoirs of depression, such as Joan Didion's *The Year of Magical Thinking,* Kay Redfield Jamison's *An Unquiet Mind* and the aforementioned William Styron's *Visible Darkness* confront mental illness stereotypes and stigmas by providing an alternative, truer story (Kramer, 2005). Through mood memoirs, individuals are invited to externalize depression as a problem to be faced and overcome, instead of an innate character flaw or unavoidable life sentence.

Rhetoric scholar Katie Rose Pryal describes the "mood memoir" as a literary genre that provides a space for those with mood disorders, who may be otherwise deemed as illegitimate rhetorical sources, to gain power through telling their stories. She argues that, "mood memoirs can be read as narrative-based responses to rhetorical exclusion by the psychiatrically disabled" (Pyral, 2010). Pyral's work outlines a new genre; however her point about gaining authority through the written, published word is salient to the idea of empowerment through creating one's own story.

Conclusion

This paper has attempted to establish that the term remission in the diagnostic criteria for major depressive disorder is problematic for numerous reasons including its attempt to apply an empirical standard to an unobservable mood disorder, as well as the potentially destructive implication that an individual can never fully recover from depression. Narrative theory aids in the examination of the DSM-5 by providing a lens through which to view the linguistic repercussions of the text. Not only does narrative theory provide a mode to critique the diagnosis, it also offers ways to use language to heal an individual diagnosed with major depressive disorder.

Narrative theory contends that language provides a system for us to make meaning of our experiences. If we are told by a medical professional with privileged vocabulary that we cannot recover from a painful disorder, but instead that we will live in a limbo vacillating between full and partial remission, we may internalize that story to become our own fate. In order to counteract this loss of agency at the hands of misused language in the DSM-5, narrative therapy and mood memoirs provide empowering autobiographical counter narratives to the stories about mental illness offered by doctors, policy makers, therapists, and the general public.

References

American Psychological Association. (2013). *Diagnostic and statistical manual of mental disorders* (5th ed.). Arlington, VA: American Psychiatric Publishing.

Ballesteros, J., Bobes, J., Bulbena, A., Luque, A., Dal-Re, R., Ibarra, N., Guemes, I. (2007). Sensitive to change, discriminative performance, and cutoff criteria to define remission for embedded short scales of the Hamilton depression rating scale (HAMD). *Jounal of Affective Disorders, (102)* Pp. 93-99.

Crawford, P., Johnson, A., Brown, B., & Nolan, P. (1999) The language of mental health nursing reports: firing paper bullets? *Journal of Advanced Nursing, 29*(2), Pp. 331-340.

Coady, N., & Lehmann, P. (2008). Theoretical perspectives for direct social work practice: A generalist-eclectic approach. New York: Springer.

Duckworth, K., Shelton, R. (2012). Depression. *NAMI, the National Alliance on Mental Illness.* Arlington, Va: NAMI.

Frank, E., Prien, R., Jarrett, R., Keller, M., Kupfer, D., Lavori, P., Rush, J., Weissman, M. (1991) Conceptualization and rationale for consensus definition of terms in major depressive disorder remission, recovery, relapse and recurrence. *Arch Gen Psychiatry 48*(9) Pp. 851-855.

Ishibashi, N. (2011) Barrier or bridge? The language of diagnosis in clinical social work. *Smith College Studies in Social Work 75* (1).

Kramer, P. (2005) "The Anatomy of Grief. Does Didion's memoir do for grief what Styron's did for depression?" Retrieved 7/1/14: http://www.slate.com/articles/arts/books/2005/10/the_anatomy_of_grief.2.html

Levitt, H., Korman, Y. & Angus, L. (2000) A metaphor analysis in treatments of depression: metaphor as a marker of change. *Counselling Psychology Quarterly 13*(1) Pp. 23-25.

Pryal, Katie Rose Guest. (2010) The Genre of the Mood Memoir and the *Ethos* of Psychiatric Disability. *Rhetoric Society Quarterly 40*(5).

Styron, W. (1990) *Darkness visible: A memoir of madness*. New York: Random House.

Workers, N. A. (2008). NASW Code of Ethics (Guide to the Everyday Professional Conduct of Social Workers). Washington, DC: NASW.

Zimmerman, M., McGlinchey, J.B., Posternak, M.A., Friedman, M., Attiullah, N., Boerescu, D. (2006). How Should Remission from Depression be Defined? The Depressed Patient's Perspective. *AM J Psychology 163*(1).

Finding the Limits with Brief Psychotic Disorder: A Social Darwinist View

Matthew Klestinski

Introduction

The objective of this essay is to present how Brief Psychotic Disorder (298.8) can be viewed through the lens of Social Darwinism as a process which reveals the personal psychic limits of an individual. This essay will argue that instead of viewing Brief Psychotic Disorder as an illness or pathology, it rather should be considered a survivable battle scar in the war of natural selection. The contributions that Social Darwinism can make to shed light on the evolving situation will be discussed. The pressures imparted by the "double bind" in social research, and the path of natural selection behind the predisposition to psychosis, will also be explored.

Brief Psychotic Disorder

To begin by drawing definitions from the DSM, the source of guidelines for diagnosis, the essential feature of Brief Psychotic Disorder is, "a disturbance that involves the sudden onset of at least one of the following positive psychotic symptoms: delusions, hallucinations, disorganized speech (e.g., frequent derailment or incoherence), or grossly abnormal psychomotor behavior, including catatonia."(American Psychiatric Association, 2013, p. 94). As there are cultural and religious backgrounds that consider such symptoms not to be aberrant, there must be a determination made whether the symptom should rather be considered a culturally sanctioned response; if so, it is not included as a symptom. Other diagnostic criteria include the following: "At least one of the symptoms must be one of the first three kinds (delusions, hallucinations, or disorganized speech) to qualify diagnostically."(p. 94) "The duration of the episode of the disturbance is at least 1 day but less than 1 month, with eventual full return to premorbid level of functioning. The disturbance is not better explained by major depressive or bipolar disorder with psychotic features or another psychotic disorder such as schizophrenia or catatonia, and is not attributable to the physiological effects of a substance (e.g., a drug of abuse, a medication) or another medical condition" (p. 94). There are three basic forms of Brief Psychotic Disorder; for the purposes of this argument, the occurrences of Brief Psychotic Disorder that occur without marked stressors, "events that, singly or together, would be markedly stressful to almost anyone in similar circumstance in the individual's culture" (p. 94) will be considered to represent more vulnerable individuals, which will be discussed later with Social Darwinism. The other two forms have diagnostic criteria specifying whether the "symptoms occur in response to marked stressors, or with postpartum onset (during pregnancy or within 4 weeks postpartum)" (p. 94) which is a stressful and significantly taxing event that is also normal in society; Brief Psychotic Disorder can also manifest with Catatonia.

At times of significant stress, Brief Psychotic Disorder may manifest. "Brief Psychotic Disorder may appear in adolescence or early adulthood, and onset can occur across the lifespan, with the average age at onset being the mid 30's." (p. 95) Temperament is a prognostic factor; "preexisting personality disorders and traits (e.g., schizotypal personality disorder, borderline personality disorder, or traits in the psychoticism domain, such as perceptual dysregulation, and the negative affectivity domain, such as suspiciousness) may predispose the individual to the development of the disorder" (p. 95).

Much as a nut may crack in the jaws of a nutcracker, so may one's psychic integrity lapse under the pressure of psychosocial stressors. Individuals with Brief Psychotic Disorder "typically experience emotional turmoil or overwhelming confusion. They may have rapid shifts from one intense affect to another" (p. 95). What could be causing these shifts? One common factor to nearly all individuals is their membership in a family. It is as part of a family that individuals first learn their roles and the rules of communication; when communication is disturbed in a family, individuals may not be taught how to effectively relieve psychic pressures.

Although Brief Psychotic Disorder and the full diagnosis of Schizophrenia are not quite the same, a similarity exists that welcomes a comparison when considering the onset of both. While exploring the psychotic condition of Schizophrenia, Gregory Bateson and his colleagues looked at how patterns of disturbed communication in a family, recurrent in similar ways in different families, resulted in very high pressure being felt by one or more members of that family system. In the process of analyzing the messages, it was found that the contradictory messages resulted in the 'victim' being placed in a no-win situation where they were in trouble if they did comply, and in trouble if they did not comply, a "double bind". If unable to withdraw from the situation, this situation would, over time, provide the ideal conditions for a psychotic break (Bateson, et.al., 1956, p. 251-254).

Social Darwinism

Darwin (1859) proposed, "..one general law, leading to the advancement of all organic beings, namely multiply, vary, let the strongest live and the weakest die" (p.263). This would be the method of selection applied by nature. Applying this to society, Social Darwinism addresses a slow and gradual evolution through natural selection into an increasingly fit mankind. Fitness would be a measure of appropriate adaptation to conditions in the eternal flux of the world. Although this selection process may be ongoing for other species in the world, this does not tend to occur so starkly within human societies who show altruism and provide social safety nets. Some who took up Social Darwinism in the past used Darwin's proposition as justification for eugenics and racism. Having Darwinism as a justification to cull those they deemed "unfit," they chose not to see that genetics alone are not the answer, but rather should be considered as part of the equation which includes the environment, that importantly includes other human beings in society, and what those humans will do when threatened. A mentally inflexible stance about insisting upon the culling of the inadequately adaptable pulls the lens of societal displeasure back to the would-be culler. This does not end well, historically.

Hawkins (1997) wrote, "Darwinism is a political theory about how new species are formed and how existing ones can become extinct (p. 24)." Hawkins went on to say, "Darwin contended that useful variations would be inherited by descendants, and the cumulative effects of this process would enable the organisms involved to mutate into new varieties, species or even genera (p. 25)." As mankind is the only species that invents and continues to adapt scientific theories of how the world works, through iterative experiment, refinement, and encouraging the plurality of diversity of ideas, it can be seen that this useful variation might be an adaptive advantage.

Hofstadter (1992) wrote about Darwin and Spencer, "An age of rapid and striking economic change, the age during which Darwin's and Spencer's ideas were popularized in the United States was also one in which the prevailing political mood was conservative (p. 5)." Hofstadter went on to say, "While not limiting the term to any 'technical' meaning, the present study likewise focuses on the specifically Darwinian concepts of struggle for existence, natural selection, and survival of the fittest – the latter Spencer's contribution, which Darwin accepted (p. 7)." Darwin was not a lone contributor to Social Darwinism, and the concepts found support as well as debate as they slowly propagated through the conservative cultural mood.

Social Darwinism in relation to Brief Psychotic Disorder

Social pressure on the macro scale tests individuals in society for vulnerability, and selects out the least adaptive. Those on the edge who can clamber back over the brink when it is discovered, sometimes with assistance, may experience personal growth as part of the struggle. When under significant mental and emotional strain, it is not uncommon to experience new insights as one adapts to the situation. But adaptation has its limits. Rarely, the would-be visionary may seek out these conditions, to test one's limits. More often, the individual arrives at this point involuntarily.

Human beings are gregarious, and tend to seek help from those they know or have easy access to in society. When it is precisely those people one would seek aid from who are providing the psychosocial pressures which are causing distress, the situation can seem intolerable. Those individuals who came from environments which taught them the fewest or least effective coping mechanisms, or, worse, taught them that it was normal to be placed in a double bind with no acceptable response, are vulnerable to entering Brief Psychotic Disorder even before being subjected to stressors or circumstances at a level that would tax anyone from that culture. These vulnerable individuals may also have the traits that would predispose them to Brief Psychotic Disorder. An individual who is suspicious of offers of help, or simply does not care for others, is much less likely to engage in the supportive care available in their area; they are exhibiting the least adaptive traits, and could be considered most at risk.

Social Darwinism has lost most of its proponents in the modern age. Most of those who were interested in the idea of evolutionary improvement through natural selection chose to use it to justify why they did not wish to aid, or wished to persecute, some other group or individuals. Even though Social Darwinism is based at the root on the idea that natural selection would slowly favor the best adaptations, no one applied this concept to mankind evolving competing societies, and societies then evolving to benefit their members. The notion that mankind, with the abilities of spoken and written communication, developing societies with diversity of thought and culture, might be able to create abundance in excess of pure survival requirements, and then use that abundance altruistically to maintain a formidable array of personalities and intellects to work on the many problems of mankind, was not the focus of the Social Darwinists of the day. But the results have become a part of modern life. Brief Psychotic Disorder is evidence of people being overwhelmed, confused, or passing their psychic limits in some way, and coming back to a full level of functioning. Western society has evolved an array of medications, quiet, safe places, and trained professionals who can identify and help individuals who are experiencing the effects of the psychosis return to symptom-free function in under a month. In developing countries, according to the DSM, it more frequently can take 1-6 months to return to a full level of functioning (American Psychiatric Association, 2013, p. 95). There is no single cause of schizophrenia yet found, so it cannot be proven that Brief Psychotic Disorder would represent those individuals who were exposed to the stressor conditions that would cause schizophrenia, but who found a way to escape the conditions, or some form of help, and in the respite, returned from the brink. There are cultures in which symptoms of Brief Psychotic Disorder, such as non-persistently hearing voices, are considered acceptable within the community. There are also religious communities which accept glossolalia as accepted parts of their community experience. If limited exposure to boundary conditions of sanity, and adaptation to that exposure, strengthens an individual, then natural selection would favor this.

It is possible that there is some hidden variable which confounds all investigation. For example, individuals who secrete a lower quantity of testosterone, a hormone which increases assertiveness, may have simply not been exposed to enough of it in their developmental years to push individuation of self to the point where the psyche is felt to be secure in the face of external opposition in adolescence and beyond. Women, with lower levels of testosterone secretion, would naturally have a greater risk for this condition than men. As of yet, Western society does not test and track hormonal levels in the citizenry to be able to make that determination.

Perhaps the true driver towards the uniquely human condition of schizophrenia is evolutionary pressure. Did natural selection through sex selection increase the likelihood of psychosis arising in humans? What if the increase in the size and function of the cerebral structures of the brain, driven by the increasing importance of language, also contributed to the predisposition for psychosis? Crow (1995) offered in his conclusions,

"In summary, key features of the theory are that the psychoses are disorders of specifically human evolution, arising from variation in the genes controlling hemispheric asymmetry that has led, by the mechanism of sexual selection, through progressive delay in maturation (neoteny) to increased brain size and intelligence. The most readily testable prediction is that the gene for asymmetry (and by implication contributing to predisposition to psychosis) should be X-Y homologous."

If the gene for hemispheric cerebral asymmetry, contributing to predisposition to psychosis, is indeed X-Y homologous, we should see twice the occurrence in females than in males. The DSM does in fact state, "Brief psychotic disorder is twofold more common in females than in males (p. 95)". This correlation is not definitive, but it does inspire interest for further investigation.

Conclusion

Social Darwinism, although mostly neglected in recent years, can still provide some illumination to the ways in which Brief Psychotic

Disorders come about. There have been social pressures on individuals, with double binds in relationship, usually familial, inducing powerful psychic pressures, that may be prerequisite significant stressors to manifest Brief Psychotic Disorders. There is a correlation, possibly worth investigating, to determine if the gene for hemispheric cerebral asymmetry in humans is X-Y homologous, leading to test Crow's theory that psychoses are disorders of specifically human evolution. Although Social Darwinism has been associated with causes that seek reasons to despise and cull the least fit, according to their own chosen definitions of fitness, the rest of the law of natural selection was ignored: those adaptations which provide an edge in fitness in the competition to evolve faster, whether it be in physical combat or sexual selection, will tend to increase in distribution throughout the population in relation to how successfully that mutation or adaptation promotes the fitness of those who carry it. As that effect of evolution through natural selection becomes more apparent, it becomes more understandable that mutations that provide sizable gains in attributes that improve the likelihood of survival of one's genetic line, will find their way into the mix. Even when those mutations occasionally have negative consequences for some offspring, the other advantages should outweigh the negatives often enough that there will be scarred survivors rather than missing descendants.

These should be considered as possible explanations for interpreting Brief Psychotic Disorder, not definitive. Even the DSM is a source of guidelines for diagnosis, and subject to revision as new research causes the body of knowledge to evolve over time. This essay has been an effort to show that humans will continue to push the limits, to thrive and learn more, and will seek brief respite and recovery when the new knowledge or implications overwhelm. Mankind will continue to learn from its mistakes and trials, and the battle scars of survived adversity will be as trophies of hard lessons learned rather than nosological signs. The ability to adapt better to changing conditions, slightly faster than a competitor, and thereby display greater fitness is probably connected to surviving overwhelm from deciding which changes in the environment are meaningful. Finding out how to best accomplish that, is the evolving puzzle Darwin unearthed.

References

American Psychiatric Association. (2013) Diagnostic and statistical manual of mental disorders (5th ed.). Washington, DC: American Psychiatric Association

Bateson, G., Jackson, D. D., Haley, J., Weakland, J., (1956). Toward a Theory of Schizophrenia, *Behavioral Science 1(4):* 251-254.

Crow, T. J., (July,1995) A Darwinian approach to the origins of psychosis. British Journal of Psychiatry, 167(1), 12-25

Darwin, C., (1859). On the Origin of Species by Means of Natural Selection. Ed. J. W. Burrow (Harmondsworth: Penguin, 1968)

Hawkins, M., (March 13, 1997). Social Darwinism in European and American Thought, 1860-1945: Nature as Model and Nature as Threat. United Kingdom, Cambridge: University Press

Hofstadter, R., (September 1, 1992). Social Darwinism in American Thought. University of Pennsylvania Press, Philadelphia

"Who Else Would I Turn To?"
Dissociative Identity Disorder In Childhood Sexual Abuse Survivors

Megan Liber

Find a group of children and ask them what they fear. One would assume common answers would include the dark, monsters, and being left alone. However, children often fear lived experiences and real people, due to past traumas in their lives. How do they cope? One such avenue is Dissociative Identity Disorder (DID), which involves a split of identities. It numbs the pain and creates a life in which the child can safely carry on. Certain states hold specific memories, while others remain oblivious to past trauma. This is particularly true of childhood sexual abuse survivors. Someone took advantage of them, and the resulting experience was so traumatizing that alternate personality states were required to manage the enduring pain and stress. The lack of a sense of agency and a fear of potential outcomes all lead to internalization. But why does it come to such an extreme? Why are there not other outlets to process in a healthy manner? This paper will explore these questions through the lens of Social Constructionism. By examining sexual abuse in children that leads to the diagnosis of DID, the connection will be made that there is no other choice but to dissociate as society has constructed a common practice of victim-blaming, leaving many in silence and shame rather than encouraging them to speak out.

Dissociative Identity Disorder

The DSM-5's diagnostic criteria for Dissociative Identity Disorder include:

Disruption of identity characterized by two or more distinct personality states...The disruption in identity involves marked discontinuity in sense of self and sense of agency, accompanied by related alterations in affect, behavior, consciousness, memory, perception, cognition, and/or sensory-motor functioning...Recurrent gaps in the recall of everyday events, important personal information, and/or traumatic events that are inconsistent with ordinary forgetting (American Psychiatric Association, 292).

The mainstays of the DID diagnosis are more than one identity and amnesia that impacts a person's daily life. Further, such marked impairment typically grows from severe trauma. "Dissociative identity disorder is associated with overwhelming experiences, traumatic events, and/or abuse occurring in childhood" (American Psychiatric Association, 294). At a time in development when safety should be guaranteed by the adults in one's life, a child's vulnerability is already high. When trauma occurs, a young person is left to navigate distress before he or she knows how. Richard P. Kluft asserts that "DID is the delivery and maintenance system...the purpose of which is the patient's internal effort to carry on with the semblance of a normal life when contending with life's actual realities has proven impossible" (2005, p. 634). When the world is no longer a safe place or becomes too overwhelming with no relief in sight, a person is left with few options to cope. A child develops his or her own set of skills through dissociating, which without proper intervention and support, will follow him or her into adulthood and impair his or her ability to function.

Childhood Sexual Abuse and Dissociative Identity Disorder

As previously mentioned, Dissociative Identity Disorder is frequently the result of trauma in which a child is left with few other options but to split into altered states to improve his or her daily functioning. "The central component of most theories that explain DID is the protective reaction to severe childhood trauma, which is often sexual in nature" (Pais, 2009, p. 73). A child who faces horrific experiences must rely on coping skills of his or her own creation. This can lead to the development of alternate identities that process and compartmentalize the memories and emotions of the sexual abuse, while others remain in an amnesic state. Kluft writes:

> In an often masochistic fashion, these identities may live within the 'terms of endearment' that prevailed in the child's dysfunctional childhood circumstances — that is, concepts of safety and goodness are understood in terms of what best (or what they thought best) protected them from further danger and harm" (2005, p. 634).

Each identity represents a personality that was found to be the safest by the child. Like a chameleon, the child assessed the situation and each state was deemed to be the best camouflage, though perhaps not consciously.

Unfortunately, such coping mechanisms set an unhealthy trend in functioning. "Events that reinforce age-specific fantasies of danger may set in motion intense anxiety and rigid early defenses, distorting the course of development" (Chertoff, 2009, p. 1427). With active imaginations, children envision the scariest of outcomes and boogie men. A quick check of the closet can reveal that no one was hiding inside. However, when real danger takes place, be it witnessed or experienced firsthand, a child's fear is strengthened and solidifies the belief in a terrifying world. In Phanicrat and Townshend's study on the coping skills of survivors of sexual abuse (2010), they found "avoidant coping strategies" to recur throughout their participants (p. 73). This avoidance leads to compartmentalizing, which in turn can lead to a split in identities the brain deems best able to handle different realities and experiences. "Sexually abused children commonly use dissociation of affect to protect themselves from overwhelming emotions, thoughts, and sensations, thus decreasing awareness of their abusive circumstances" (Somer & Szwarcberg, 2001, p. 333). Rather than growing with guidance from adults who help them to understand and process their feelings, a crude, immediate emergency response takes place. In order to survive, a child adopts a Dissociative Identity Disorder.

Social Constructionism

The theory of Social Constructionism views reality as one that has been created by the whole social group. These constructed ideals become guidelines for all of society, which in turn become indoctrinated within the individual. Owen describes this as "the claim and viewpoint that the consent of our consciousness, and the mode of relating we have to others, is taught by our culture and society: all the metaphysical quantities we take for granted are learned from others around us" (1992, para. 5). Unconsciously, we are creating rules and ideas about how the world works. Commonly-held beliefs found in society are continuously perpetuated without a thought as to why. "In the case of social constructionism it is the denial of subjectivity and agency, of the 'I' the speaker has, but the socially produced person apparently has not" (Craib, 1997, p. 5). In a densely populated world, it would be rare for a person to grow up outside of social influence and some form of community. A child takes in messages from the behavior and reactions of the people around him or her. The child begins to form a sense of what is right and wrong. Then he or she follows along without realizing why. Members of society, in the viewpoint of social constructionism, do not see "[t]he taken for granted assumptions...includ[ing] norms or appropriate, and abnormal or inappropriate behaviour, thought and emotion--by considering who is acting on whom and in what context (Owen, 1992, para. 11). Quite often a person believes something to be true because it is popular opinion. Why go against the grain and think differently, particularly in the minute details of the way the collective observes and thinks about one another? In Owen's description:

> People are not individuals in the sense that they are alone and isolated. They are points in a flux or are single places of resonance within a field of human forces which pull and push the participants in different directions. People are like ants in an ant colony or like the legs on a milli-pede that move in step with each other. Furthermore we cannot be individuals in the narrow sense of being distant and separable from others. We continually take in the in-fluences of others and send out messages of how to be that are taken in by others. In this sense it is impossible to strive to be an individual because you can never be one: you can only swim with or against the tide of others (1992, para. 14).

Social Constructionism foretells a world in which one finds it dif-ficult to fight against the collective and its beliefs. Go to the store and take a look at baby clothes. Why do we associate pink and purple with girls and blue and green with boys? It may be safe to assume that most people could not think of a logical, concrete explanation. Quite simply, it comes down to the plain fact that somewhere along the line the group as a whole constructed the idea that such colors connote each specific gender, and this idea then influences baby an-nouncements and showers, clothes, toys, and media. Attempting to go against such societal decisions may prove to be difficult.

A major component of Social Constructionism is the use of lan-guage to define society and our interactions. For example, we know Dissociative Identity Disorder as such because we have chosen to de-fine it and give it meaning. "[A] social constructionist view promotes a Copernican shift in our understanding of language, from seeing it as a medium for merely reflecting or labeling an independent reality, to viewing it as the very medium by which social reality is constructed" (Neimeyer, 1998, p. 139). From this viewpoint, it becomes of the utmost importance to question the power given to certain words and ideas that hold stock. Who deemed this to be so? Why do we believe what we do? Somewhere along the line, those with darker skin were deemed inferior and suffered years of slavery and discrimination. In the end, socially constructed, archaic ideals can lead to bondage, both large and small.

Victim-Blaming Culture

Particularly in regard to sexual crimes, victims are frequently blamed for the crime. For example, when a rape victim comes forward to report what happened, his or her character is often called into question. "Victims who behave in a manner perceived by others to be atypical may be perceived as more blameworthy" (Fox & Cook, 2011, p. 3409). What were they wearing? Were they being a "tease?" Have they been intimate with this person before? Were they drunk? Somer and Szwarcberg describe this pervasive thinking:

> Prevailing attitudes toward rape and rape victims are often mirrored in the legal and judicial systems. Survivors of sexual assault frequently complain that their experience with the criminal justice system was humiliating. Inappropriate sexual interest, derogatory questioning, and a generally disrespectful attitude often mark the process (2001, p. 334).

A common conception is, "If I am in danger, the police will protect me." However, where does a rape victim go if they are the ones on trial? After going through a humiliating, traumatizing experience, they are forced to further endure probing questions that may not have been asked if they were simply robbed on the street.

The perpetrator's future and reputation are often seen as fragile. The media gives news reports on the damage such charges will bring to the assailant's life. Those who report a sexual assault can be seen as selfish, silly people who got in too deep and then misconstrued the situation into a giant mess. Taken a step further, Johnson, Mullick, and Mulford assert:

> In addition, individuals may be more likely to blame the victim if they hold strongly to a *belief in a just world,* according to which the world is largely a safe and fair place in which to live. Those who believe in a just world may be more inclined to believe that 'good things usually happen to good people' and that 'bad things usually happen to bad people.' Thus, the judging individual may assume that, if others are in trouble, then their problems are likely to be their own fault (2002, p. 250).

It becomes the belief that the persons brought this on themselves. This idea seems to stem back to religious morality and the notion that those who find themselves in trouble must have been leading a sinful life that led up to this point. Those who lead pure lives will be rewarded. Just as the Victorian Poor Laws decided who the worthy and unworthy poor were, so do we today in our assessment of victims of sexual assault.

An assumption could be made that since children are small and vulnerable, society is less likely to question sexual abuse claims. Unfortunately, victim-blaming is interwoven in the fabric of our communities such that even a child cannot escape it. In their article on the psychological effects of childhood abuse in women, Briere and Jordan state:

[S]ocially transmitted attitudes appear to increase the likelihood of child abuse. These include myths about children's sexuality, especially that children may desire sexual contact with parents or other adults and thus may behave in ways that lead to their own victimization; beliefs regarding the social appropriateness of corporal punishment, which may extend to behaviors that meet current definitions of physical child abuse; and broader-based support for parents' rights over those of children, including acceptance of parental authoritarianism. These various attitudes and beliefs are often communicated to the child, who may, as a result, feel responsible for maltreatment (i.e., because she 'asked for it' or deserved it) or believe that she is not entitled to better parental behavior and care. These understandings, in turn, may lead to guilt, shame, or denial, all of which may compound or exacerbate abuse effects (2009, pp. 379-380).

Assuming a child has sexual urges and desires only enables the victim-blaming. Adults may know best about when to eat dessert, stranger danger, and proper bedtimes, but what happens when guidance becomes confused with discipline and control? The well-being of the child becomes lost. As they find their parents to be trustworthy, doubt and self-blame invade the child's view of him or herself. A pattern has been created that leads a child to believe that the world is a cruel place.

Analysis

When looking at victim-blaming through the theory of Social Constructionism, it becomes a created societal mode of behavior. Unfortunately, it leads a person to believe, either through the language he or she hears from others or personal or observed experience, that disclosing what happened could lead to further harm. Somer and Szwarcberg found for childhood survivors of sexual abuse, "[a]mong factors that contributed prominently to the delayed disclosure were fears of social rejection and condemnation, mistrust of people, and adoption of family-indoctrinated values of obedience" (2001, p. 338). Even a child recognizes the implications socially if he or she chooses to share what happened. After assessing the potential for victim-blaming and support, children dissociate into other identities in order to function and survive. Adults do not always provide respect to the young, seeing them as immature, imaginative, and unreliable. They receive blame for making things up, misunderstanding, being dramatic, and for their emotionality. "Children do not get it, so how can we take their accusations seriously?" Yes, kids do have fallible memories at times, but is that to say that nothing a child shares or believes in is real? It becomes a slippery slope to argue strongly in such a direction.

The idea of family brings to mind loyalty, love, and support. Social Constructionism sees society as the entity that developed the idea that families stick together and protect one another. Yet, by the same token, one would never air the group's dirty laundry. Keep secrets, suffer in silence, and maintain a picture perfect image. "Child victims of sexual abuse, particularly intrafamilial abuse, may not be willing to sacrifice the integrity of their families and their sense of belonging to it" (Somer & Szwarcberg, 2001, p. 339). In order to maintain what collective society deems important, children hide their pain. If families do find out, they use victim-blaming in order to conform to socially constructed norms.. "Shame among family members is...common. It may take the form of urging the survivor to keep quiet about the assault and pretend it never happened or outright disgracing her for having been violated" (Moor, 2007, p. 23). If a child was not aware of it before, he or she quickly recognizes the power of victim-blaming. The shame and guilt can damage his or her self-worth, and the re-victimization can reinforce Dissociative Identity Disorder.

Conservative homes often adopt a tight-lipped stance on communicating within the family about sex. Children may not know what good touch, bad touch is, or even anything about their private parts for that matter. "The dynamics of sin and shame, and the lack of family idioms with which to discuss sexual behavior, characteristics common to very conservative communities, could contribute to CSA [childhood sexual assault] victims' difficulties in disclosing their plight" (Somer and Szwarcberg, 2001, p. 334). Lack of information and perhaps learning only that their private parts are sinful, opens the door to both self-blame, and victim-blaming from the adults who view them as impure. Children learn that "if adults do not talk about it, then I do not talk about it."

Just as adult sexual assault victims' choices and behavior are called into question, so do children face specific victim-blaming judgment. Fox and Cook point out:

> Prior research suggests that both female and male children who resisted sexual abuse were least likely to be blamed, while children perceived as physically encouraging the abuse were the most likely to be blamed and were perceived to have suffered less harm (2011, p. 3409).

Victim-blaming is so indoctrinated throughout society that even a child, who is most likely sexually abused by someone older and/or bigger than him or her, is culpable in his or her violated role. Either way, a child may now see the world as a justice-less place and find only him or herself to turn to in distress. How can one person provide the kind of support necessary to overcome trauma? By splitting into alternate states that are deemed useful and soothing in the brain.

Victim-blaming inhibits childhood sexual abuse survivors from receiving support. Phanicrat and Townshend found in their study on coping strategies for those who suffered from childhood sexual abuse "that seeking support is an important step on the road to recovery as survivors are likely to receive appropriate psychological interventions that subsequently enhance effective coping skills by using strategies such as cognitive reframing, self-reflection, and positive thinking" (2010, p. 74). In order to prevent the development of Dissociative Identity Disorder, children must be reached early with proper aid in the healing process. By listening, validating, acquiring therapy, and getting out of the dangerous situation, CSA survivors can move forward with the development of healthy coping skills. Victim-blaming only leads to dysfunctionality that will impact their functioning well into adulthood.

Perhaps the time has come to turn it back on those with victim-blaming tendencies. As a method of coping, some children are merely leaning on Dissociative Identity Disorder because of the dysfunction of others. Though it cannot be known for certain and would require further research, one wonders if the diagnosis would cease to exist in a world without societally constructed victim-blaming. Such a belief-system shows a lack of empathy in those quick to point fingers, and an unnecessary coldness. Maybe the DSM requires an additional diagnosis under the personality disorders: Apathetic Personality Disorder. Should not this type of behavior be of more concern?

Conclusion

Dissociative Identity Disorder in childhood sexual abuse survivors remains prevalent due to societally constructed victim-blaming. Negative messages, language, and interactions observed by children can lead them to a life of suffering in the dark. Only in the darkness does a child find his or her own methods of coping. The brain decides, "If I cannot find a real person to help me, I will create one to make all of this manageable." Amnesia and altered states come forward to comfort the child and allow for survival. However, this leaves a disturbing trend that begs the question, when will society construct a new methodology? When will enough be enough? We teach children stranger-danger, but when will that become just plain danger-danger? The time has come for a new formulation of ideals in order to create a victim-centered approach. Not just for children, but for all ages and stages. Perhaps a reduction will be seen not just in the cases of Dissociative Identity Disorder, but in other diagnoses as well. If the DSM recognized the impact of an apathetic-type personality disorder, the world may see a reduction in the diagnosis of DID. Through proper therapy, this new societal construct would hopefully be achieved. A world would exist in which all thoughts and feelings would be shared, and childhood sexual abuse victims would be awaiting outstretched, open arms.

Unfortunately, until that time arises, CSA survivors will continue to be shamed by society. Victim-blaming will carry on. Until a new social construction enables victim-supporting, the harshest of abusers may never learn unless they experience it for themselves. Dissociative Identity Disorder will continue to flourish in the minds of many until the day comes when their choice of healthy coping skills are plenty.

References

American Psychiatric Association. (2013). *Diagnostic and Statistical Manual of Mental Disorders* (5th ed.). Washington, D.C.: American Psychiatric Publishing.

Briere, J., & Jordan, C. (2009). Childhood Maltreatment, Intervening Variables, and Adult Psychological Difficulties in Women. *Trauma, Violence & Abuse: a Review Journal*, 10.4, 375-388.
http://tva.sagepub.com.flagship.luc.edu/content/10/4/375.full.pdf+html

Chertoff, J. M. (2009). The complex nature of exposure to early childhood trauma in the psychoanalysis of a child. *Journal of the American Psychoanalytic Association*, 57.6, 1425-1457.
http://apa.sagepub.com.flagship.luc.edu/content/57/6/1425.full.pdf+html

Craib, I. (1997). Social Constructionism as a Social Psychosis. *Sociology*, 31.1, 1-15. http://soc.sagepub.com.flagship.luc.edu/content/31/1/1.full.pdf+html

Fox, K. A., & Cook, C. L. (2011). Is knowledge power? The effects of a victimology course on victim blaming. *Journal of Interpersonal Violence*, 26.17, 3407-27. http://jiv.sagepub.com.flagship.luc.edu/content/26/17/3407.full.pdf+html

Johnson, L. M., Mullick, R., & Mulford, C. L. (2002). General versus specific victim blaming. *The Journal of Social Psychology*, 142.2, 249-263. http://web.a.ebscohost.com.flagship.luc.edu/ehost/detail?sid=e59bdbad-035d-40cd-b317-ee0af26a878c%40

Kluft, R. P. (2005). Diagnosing Dissociative Identity Disorder. *Psychiatric Annals*, 35.8, 633-647. http://web.b.ebscohost.com.flagship.luc.edu/ehost/pdfviewer/pdfviewer?sid=b5a0a3c4-d8f1-48ea-bf93-76fd96346a24%40sessionmgr111&vid=2&hid=117

Moor, A. (2007). When Recounting the Traumatic Memories is not Enough: Treating Persistent Self-Devaluation Associated with Rape and Victim-Blaming Rape Myths. *Women & Therapy*, 30.1-2, 19-33. http://www.tandfonline.com.flagship.luc.edu/doi/citedby/10.1300/J015v30n01_02#tabModule

Neimeyer, R. A. (1998). Social constructionism in the counselling context. *Counselling Psychology Quarterly*, 11.2, 135-149. http://web.b.ebscohost.com.flagship.luc.edu/ehost/detail?sid=22e3f99c-66a9-4ceb-83a1-5df8a68627e7%40

Owen, I. R. (1992). Applying social constructionism to psychotherapy. *Counselling Psychology Quarterly*, 5.4, n.p. http://web.a.ebscohost.com.flagship.luc.edu/ehost/detail?sid=6bfda9f3-fddf-485c-91b3-e0d089780555%40

Pais, S. (2009). A Systemic Approach to the Treatment of Dissociative Identity Disorder. *Journal of Family Psychotherapy*, 20.1, 72-88. http://www.tandfonline.com.flagship.luc.edu/doi/abs/10.1080/08975350802716566#.U7nfVV79PwI

Phanichrat, T., & Townshend, J. (2010). Coping Strategies Used by Survivors of Childhood Sexual Abuse on the Journey to Recovery. Journal of Child Sexual Abuse, 19.1, 62-78. http://web.b.cbscohost.com.flagship.luc.edu/ehost/detail?sid=3f8a1a5f-1edd-4706-8455-bff3ac6bc31c%40

Somer, E., & Szwarcberg, S. (2001). Variables in delayed disclosure of childhood sexual abuse. *The American Journal of Orthopsychiatry*, 71.3, 332-41. http://onlinelibrary.wiley.com.flagship.luc.edu/store/10.1037/0002-9432.71.3.332/asset/0002-9432.71.3.332.pdf?v=1&t=hxas8h3c&s=9d9f01818e1b75136a33a870e46ce4612c7b68ff

The Tie That Binds: A Bowen Family Systems Perspective on Anxiety Disorders

Angela McLean

Anxiety and fear are normal reactions that can serve as protective measures during threatening situations. However, an anxiety disorder may be diagnosed when symptoms are experienced for an extended period of time and impair the functioning in one's daily life (American Psychiatric Association [APA], 2013). This diagnostic category includes twelve different diagnoses, which "differ from one another in the types of objects or situations that induce fear, anxiety or avoidance behavior" (APA, 2013, p. 189). The symptoms include difficulty separating from home, persistent fear of specific objects, paralyzing fear in social situations as well as somatic symptoms such as sweating and heart palpitations (APA, 2013). According to the National Institute of Mental Health (NIMH), anxiety disorders are the most common mental health disorders in the United States, impacting more than 40 million adults, (NIMH, 2009). It is acknowledged in the Diagnostic and Statistical Manual of Mental Disorders (5th ed.; DSM-5) that several anxiety disorders develop during childhood and often endure later in life if not treated (APA, 2013). Why and how do many of these disorders first appear during childhood? The DSM-5 recognizes environmental and genetic factors when assessing risk for anxiety disorders, but how do these variables intersect when making a diagnosis?

According to Bowen family systems theory "each of us is best understood not as an individual psychological unit, but as a functional part of an emotional unit – the family" (Bowen, 1978, p.188). Through this lens, mental health issues are not developed from within the individual but rather as a result of the interactions that take place in the family environment. Therefore, a person never truly functions on his or her own, as the family is an intensely connected, emotional unit. Dr. Murray Bowen developed this theory based on systemic thinking to incorporate the interconnectedness of human families and its transmission of anxiety from one member to another (Bowen, 1978). Bowen family systems theory is based on eight, interrelated core concepts. Among these concepts are differentiation of self, family projection process, nuclear family emotional system, triangulation, multigenerational transmission process and sibling position.

Differentiation of Self

Differentiation of self refers to an individual's ability to balance the forces of togetherness and individuality within one's family. "Feelings are often assumed or shared between family members due to unconscious reactivity that has become a response pattern within the system" (Kolbert, Crothers & Field, 2013, p. 87). Individuals who are less differentiated demonstrate emotional fusion, also known as interdependent thinking and feeling. They hold others responsible for their happiness, have difficulty distinguishing between emotional and intellectual functioning and make decisions in an attempt to relieve their anxiety as well as the anxiety of others. Additionally, their decisions are driven by a need to seek approval from others (Bowen, 1978). By contrast those who are more differentiated have a self that is made of "clearly defined beliefs, convictions and life principles" (Bowen, 1978, p. 365).

Family Projection Process

According to the Bowen Center (2014) the family projection process "describes the primary way parents transmit their emotional problems to a child" (Family Projection Process, para. 1). When a parent has unresolved anxiety or a less differentiated self, he or she will project these feelings onto the child. In turn, this establishes "a pattern of infantilizing the child, who gradually becomes more impaired" (Bowen, 1978, p. 381).

Parental control is one family projection process that can contribute to a poorly differentiated self and one that is recognized as an environmental risk factor for anxiety by the DSM-5. Controlling parents exhibit "overprotection, oversolicitousness and intrusiveness" as a means to regulate their children (Hudson, Dodd & Bovopoulos, 2011, p. 940). In general, anxious children have parents who are over-involved compared to parents of children who are not anxious (Hudson et al., 2011). This is a theme that is all too familiar in modern-day parenting. Today's parents can be overly domineering in many aspects of childrearing, frequently robbing young children of experiences that are vital to their developing autonomy. The abundance of parental control is apparent in the number of child "leashes" seen at zoos across the United States. Children are often not given the opportunity to freely explore their environment or resolve conflicts with peers without adult intervention. This excess of parental involvement can leave children feeling anxious about new experiences and apprehensive of their abilities to handle situations independently.

The link between parenting style and the development of anxiety disorders later in life is the most widely studied family connection in this category (Rapee, 2012). A variety of studies, including observational and qualitative methods have shown links between parental overprotection and child anxiety (Rapee, 1997). Two of these studies have shown links between anxiety in children as young as preschoolers and overinvolved parenting (Bayer, Sanson & Hemphill, 2006; Edwards, Rapee & Kennedy, 2010). Additionally, a recent longitudinal study of more than 3,000 adolescents showed a significant link between anxiety disorders and parental overprotection (Beesdo, Pine, Lieb & Wittchen, 2010). According to Wood, McLeod, Sigman, Wang & Chu, "The different manifestations of parental control each involve encouragement of children's dependence on parents, which is hypothesized to affect children's perceptions of mastery over the environment" (2003, p. 135). This lack of mastery does not allow children to fully differentiate from their parents, in turn leading them to have later difficulties exhibiting independence and trusting in themselves. As these children become adolescents and young adults they are instead likely to make decisions solely based on what is pleasing to their parents. This is another phenomenon that is common today. Parental intrusiveness is robbing this up-and-coming generation of practice in social situations, leaving them unprepared for future occupational and personal endeavors. Anecdotal reports from high school and university professors tell stories of students who rely on mom or dad to dispute their grades, complete assignments and manage conflicts with friends.

Aggressive parenting is an additional family projection process that can contribute to a poorly differentiated self. Aggression is an act that is used to induce fear and submissiveness in others (Keltner and Kring, 1998). When used in parenting, aggression leads to a child who becomes and is likely to remain emotionally fused with parents. A longitudinal study of the connection between the development of anxiety and adolescent-parent interactions showed that higher levels of parental aggression resulted in increased rates of anxiety two years later (Schwartz, Dudgeon & Allen, 2012). This observational-based study is evidence of the interconnection of family dynamics and anxiety. Parents continue to have a substantial impact on adolescent development despite the growing relevance of peer relationships during this life stage (Steinberg and Morris, 2001).

Although the DSM-5 does address the development and course of specific anxiety disorders, a systemic approach would provide a much-needed comprehensive view of the variance based on developmental stages. By applying a family systems lens to diagnosis, the clinician would not only recognize internal changes that are relevant to each life stage, but also how the relationship between family members progresses as children age. According to Schwartz et al. (2012), "adolescence is a period of particular sensitivity for the development of...anxiety, and is a period in which the parent-child relationship undergoes change" (p. 61). As adolescents become more independent they are not as close to their parents as they once were, and there is a slight rise in conflict between parents and children of this age (Steinberg and Morris, 2001). As parent-child conflict increases during this developmental stage, clinicians should be mindful of this changing dynamic when assessing adolescents for anxiety. Increased social and academic demands combined with changing biochemistry make this age group especially vulnerable. Parents of adolescents should also be cognizant of the methods of control imposed on their children in this stage.

Nuclear Family Emotional System

Another concept of Bowen family systems theory, the nuclear family emotional system, describes relationship patterns that highlight where family problems develop. According to the Bowen Center (2014), "Clinical problems or symptoms usually develop during periods of heightened and prolonged family tension.The higher the tension, the more chance that symptoms will be severe and that several people will be symptomatic" (Nuclear Family Emotional System, para. 2). There are four relationship patterns where problems are likely to originate in the family system: marital conflict, compromised emotional functioning in one spouse, emotional impairment of children and emotional distance (Walsh and Harrigan, 2003).

More than 1,200 adolescents provided self-reports of parental violence and were subsequently evaluated for psychosocial outcomes, including anxiety (Fergusson and Horwood, 1998). This study found that witnessing parental conflict during childhood significantly predicted anxiety disorders in adolescence. Specifically, parental violence directed towards the mother from the father impacted anxiety outcomes, while mother-initiated acts did not. Additionally, the results mitigated previous suggestions that gender acts as a modifier, as adolescent boys and girls did not differ in their reactions to witnessing extreme marital conflict. The levels of conflict were scaled, ranging from mild (verbal name calling) to extreme (choking), and witnessing higher levels of conflict resulted in a higher likelihood for anxiety symptoms.

Another longitudinal study found that children who witnessed parental conflict as kindergarteners were more likely to experience anxiety symptoms seven years later (Cummings, George & McCoy, 2012). Marital disagreements in which parents displayed positive conflict resolution skills were not damaging to children, but those including violence, harsh criticism or stonewalling did impair psychosocial functioning. According to Cummings et al. (2012), "early experiences with family adversity prime children to be sensitive and vigilant to bouts of interparental conflict throughout childhood and adolescence" (p. 1711).

It is likely that the modeling of positive conflict resolution skills would minimize anxiety for children. This modeling would destigmatize disagreements and provide the reassuring message that people are capable of working out their problems together. Individuals of all ages instinctively prefer harmonious interactions between their family members, peers and coworkers. Children are especially vulnerable to the conflict they witness between parents, as these are typically their primary caregivers and earliest role models. Frequent parental conflicts send the message to children that the world is an unpredictable, unsettling place, and these messages are likely to invoke anxiety when exposure is prolonged. The study by Cummings et al. (2012) demonstrates that it is important to historically assess familial factors when evaluating adolescents, as conflicts witnessed several years prior are shown to have lasting impact on anxiety.

Triangulation

Heightened marital conflict impacts every member of a household. The family system is connected in such a way that each member can be drawn into disputes in various ways. Triangulation is another concept of Bowen family systems theory that is common in any intimate relationship (Bowen, 1978). The cost of close relationships is intermittent struggles, and "when in conflict people usually rely on a third person for mediation, ventilation or problem-solving assistance" (Walsh and Harrigan, 2003, p. 385).

These triangles may become maladaptive when the third person is a child on the outside of marital conflict. Anxiety is likely to be experienced as a result of this position within a family system. The DSM-5 recognizes parental divorce as an environmental risk factor for separation anxiety, but what about the martial conflict and potentially harmful triangulation patterns that are likely to precede divorce (APA, 2013)? The DSM-5 fails to consider the implications of parental conflict on anxiety development and instead places emphasis on genetic risk factors, citing genetic vulnerabilities for nearly all anxiety disorders.

Multigenerational Transmission Process

According to Bowen family systems, the concept of multigenerational transmission "defines the principle of projection of varying degrees of immaturity (undifferentiation) to different children when the process is repeated over a number of generations" (Bowen, 1978, p. 205). Family models of interaction as well as individual personality traits have origins in preceding generations (Walsh and Harrigan, 2003). A pattern of family members with low levels of differentiation is likely to equate to a higher likelihood for anxiety disorders. Bowen (1978) states that those with the lowest levels of differentiation "live in a feeling-dominated world in which it is impossible to distinguish feeling from fact" (p. 367). This is a characteristic consistent with traits of anxiety disorders, in which feelings dominate the "anticipation of future threat" (APA, 2013, p. 189).

Although the concept of multigenerational transmission is not based on a genetic model, patterns of psychopathology can be observed when assessing the family system through a genogram. A genogram is a visual tool that represents characteristics and relational patterns within three generations of a family (Walsh and Harrigan, 2003). The DSM-5 identifies genetics as a predominant risk factor for all anxiety disorders (APA, 2013). In fact, this is the only risk factor that is consistently listed among all disorders in the anxiety diagnostic category.

Undoubtedly, anxiety disorders tend to run in families. There are several studies that show higher levels of anxiety among first-degree relatives of individuals with anxiety disorders (Hettema, Neal & Kindler, 2001). However, the DSM-5 oversimplifies the family connection by attributing this to genetic factors. Research actually shows that heredity may be overestimated (Bogels, 2006). Genetic studies estimate a higher heredity of temperament traits such as behavioral inhibition (BI) or fear of negative evaluation than anxiety disorders. Therefore, the mechanism by which anxiety is transmitted is not clear. Family dynamics such as parenting style, marital conflict and parental projection of emotional problems are probable contributors to the development of anxiety and may explain the high rates of psychopathology within families. Although multigenerational patterns of anxiety are observed within families, these repetitions are likely related to repeated cycles of behaviors rather than pathology within the individuals. Further studies are needed to examine the ultimate link between environment and genetics when assessing and diagnosing anxiety disorders.

Sibling Position

Bowen (1978) incorporated the work of psychologist Walter Toman on sibling birth order into his family systems theory. The underlying concept is that those who grow up in comparable sibling positions tend to share similar characteristics. Research into the association between birth order and anxiety has produced conflicting results, but most has found children with a later sibling position to be more anxious and shy than first-born siblings (Bogels, 2006). Anxiety in younger siblings is found when older siblings dominate them and they are the recipients of negative comments from an older brother or sister. Additionally, children with anxiety believe that their parents express favoritism towards their sibling(s) more than children without anxiety (Lindhout et al., 2003).

Sibling position does not dictate anxiety, nor should younger birth order position be included as a risk factor for specific disorders. However, from a family systems perspective the interaction between siblings is an important factor when considering a diagnosis. A relevant consideration, according to Bogels (2006), is that "positive sibling relationships may protect a child against the negative influence of, for example, marital conflict" (p. 840). This observation demonstrates many aspects of Bowen family systems theory. Constructive affiliations between family members can positively impact self differentiation, reduce the negative impact of triangulation and strengthen the nuclear family emotional system.

Bowen family systems theory claims that the impact of family members on one another is so overwhelming it is as if they live under the same emotional skin (The Bowen Center, 2014). According to the Bowen Center (2014),

> When family members get anxious, the anxiety can escalate by spreading infectiously among them. As anxiety goes up, the emotional connectedness of family members

becomes more stressful than comforting. Eventually, one or more members feel overwhelmed, isolated, or out of control.

From this perspective, there is no way to make an anxiety disorder diagnosis without first considering all the nuances of family interactions. Clinicians should carefully assess the interconnected, relational patterns within families and not rely solely on presumed genetic factors. Although the DSM-5 lists genetic vulnerability as a risk factor for all anxiety disorders, there is still too much unknown about this link to assume that heredity is the answer for family transmission patterns of anxiety. Shared genes may technically make a family, but the emotional connectedness of individual members is truly the tie that binds.

References

American Psychiatric Association. (2013). *Diagnostic and statistical manual of mental disorders* (5th ed.). Arlington, VA: American Psychiatric Publishing.

Bayer, J. K., Sanson, A. V. & Hemphill, S. A. (2006). Parent influences on early childhood internalizing difficulties. *Journal of Applied Developmental Psychology*, 27, 542-559.

Beesdo, K., Pine, D. S., Lieb, R., & Wittchen, H. U. (2010). Incidence and risk patterns of anxiety and depressive disorders and categorization of generalized anxiety disorder. *Archives of General Psychiatry*, 67(1).

The Bowen Center for the Study of the Family. (2014). Retrieved from http://www.thebowencenter.org.

Bowen, M. (1978). *Family therapy in clinical practice*. New York: Aronson.

Edwards, S. L., Rapee, R. M., & Kennedy, S. (2010). Prediction of anxiety symptoms in preschool-aged children: examination of maternal and paternal perspectives. *Journal of Child Psychology and Psychiatry*, 51, 313-321.

Fergusson, D. M., & Horwood, L. J. (1998). Exposure to interparental violence in childhood and psychosocial adjustment in young adulthood. *Child Abuse and Neglect*, 32 (6).

Hettema, J. M., Neale, M. C., & Kendler, K. S. (2001). A review and meta-analysis of the genetic epidemiology of anxiety disorders. *American Journal of Psychiatry*, 158(10), 1568–1578.

Hudson, J. L., Dodd, H. F., & Bovopoulos, N. (2011). Temperment, family environment and anxiety in preschool children. *Journal of Abnormal Child Psychology*, 39, 939-951.

Keltner, D., & Kring, A. M. (1998). Emotion, social function, and psychopathology. *Review of General Psychology*, 2, 320–342.

Kolbert, J. B., Crothers, L. M., & Field, J. (2013). Clinical interventions with adolescents using a family systems approach. *The Family Journal: Counseling and Therapy for Couples and Families,* 21 (1), 87-94.

Lindhout, I. E., Boer, F., Markus, M. T., Hoogendijk, T. H. G., Maingay, R., & Borst, S. R. (2003). Sibling relationships of anxiety disordered children—A research note. *Journal of Anxiety Disorders*, 17, 593−601.

National Institute of Mental Health. (2009). *Anxiety disorders* [Booklet]. Bethesda, MD: United States Department of Health and Human Services.

Steinberg, L. & Morris, A. S. (2001). Adolescent development. *Annual Review of Psychology*, 52, 83-110.

Schwartz, O. S., Dudgeon, P., Sheeber, L. B., Yap, M. B. H., Simmons, J. G., & Allen, N. B. (2012). Parental behaviors during family interactions predict changes in depression and anxiety symptoms during adolescence. *Journal of Abnormal Child Psychology*, 40, 59-71.

Walsh, J & Harrigan, M. (2003). The termination stage in bowen's family systems theory. *Clinical Social work Journal.* 31 (4). 383 – 394.

Wood, J., McLeod, B. D., Sigman, M., Hwang, W. -C., & Chu, B. C. (2003). Parenting and childhood anxiety: Theory, empirical findings, and future directions. *Journal of Child Psychology and Psychiatry*, 44, 134−151.

Anxiety & Cognitive Behavioral Theory

Allison Riggio

Anxiety—one of the most common mental health issues in America—is among the most loosely defined disorders in the DSM-5. According to the National Institute of Mental Health, 40 million American adults have had some type of anxiety disorder, which is a startling rate. This prevalence would lead one to believe that the diagnostic criteria would be detailed and definitive. However, most of the diagnostic criteria for the anxiety disorders in the DSM-5 make little sense for actually diagnosing a client. They focus too much on what the client is experiencing and how that impacts his or her daily life; and ignore the reason the client is experiencing that anxiety in the first place. Through the view of cognitive behavioral theory, we can reasonably conclude that the root cause of the client's anxiety is the most important aspect of the therapeutic process.

The Anxiety Disorders chapter of the DSM-5 is specific on the macro level—separating out varying kinds of anxiety and specific phobias; but it is vague on the micro level—failing to list direct, clear diagnostic criteria for clinicians to use. The chapter lists the following 12 anxiety disorders, in order of typical age at onset: separation anxiety disorder, selective mutism, specific phobia, social anxiety disorder (social phobia), panic disorder, panic attack specifier, agoraphobia, generalized anxiety disorder, substance/medication-induced anxiety disorder, anxiety disorder due to another medical condition, other specified anxiety disorder, and unspecified anxiety disorder (American Psychiatric Association, 2013).

The American Psychiatric Association takes great care in categorizing the different forms anxiety can take, but lacks that same diligence when it comes to detailing each disorder. For example, the

DSM-5 uses circular definitions to describe certain diagnostic criteria. In some regards, this is acceptable because the manual was written under the assumption that the reader will have some existing knowledge and understanding of the content. But perhaps the most egregious overuse of a circular definition is the use of the word "anxiety" in the diagnostic criteria for 11 of the 12 anxiety disorders listed in this chapter—without ever defining anxiety whatsoever. Doing so is paradoxical at best, and downright confusing at worst. The role of the DSM is to be the industry standard in identifying and diagnosing disorders; yet the anxiety disorders chapter of the DSM expects readers to already know how to identify anxiety. The American Psychiatric Association would never list "excessive schizophrenia" as a diagnostic criterion for schizophrenia, because it does not tell us what it means to be schizophrenic. As such, clinicians will be no better at diagnosing anxiety by being told the required criteria for an anxiety disorder is anxiety.

In addition to the circular logic used to define anxiety, the disorders in this chapter of the DSM-5 are bursting at the seams with examples of ambiguous, subjective qualifiers. For example, the number one diagnostic criterion for generalized anxiety disorder is "excessive anxiety and worry (apprehensive expectation), occurring more days than not for at least six months, about a number of events or activities (such as work or school performance)" (American Psychiatric Association, 2013, p. 222). The use of words and phrases like "excessive," "more days than not," and "a number of events" are subjective and leave the interpretation up to the judgment of the clinician. While the clinician is indeed the expert on these matters, it is unsettling that the primary diagnostic criterion is written with loosely defined qualifiers that give no scale or scope for the client or the clinician to work from. Though, it does not seem the American Psychiatric Association cares much about the client's perspective when it comes to anxiety disorders, considering the chapter introduction states the following:

"Since individuals with anxiety disorders typically overestimate the danger in situations they fear or avoid, the primary determination of whether the fear or anxiety is excessive or out of proportion is made by the clinician, taking cultural contextual factors into account" (American Psychiatric Association, 2013, p. 189).

The various anxiety disorders in the DSM-5 have no significant difference between one another in diagnostic criteria, because the entire chapter is incredibly generic. Each disorder results in psychological distress brought about by one or more physical or emotional fears, and the only differences appear in the length and type of symptoms present. Therefore, in selecting a diagnosis, clinicians must focus solely on the effect of the anxiety, not the cause. For example, without knowing the catalyst for the anxiety, a clinician might struggle to differentiate chronic stress from generalized anxiety disorder. If clinicians are not encouraged to look at the client as a unique being with a history that has impacted the onset and severity of certain fears, then they will never properly diagnose or treat the anxiety disorder.

Cause and effect is a unique relationship in the world of psychology. The idea that there is some explanation for why we do what we do has been researched and explored for more than 75 years. Initial founders of behaviorist research focused on the cause and effect relationship from many angles. Ivan Pavlov studied the way dogs responded to stimulus by conditioning the animals to expect and anticipate food upon hearing the sound of a ringing bell. Another father of behaviorism, B.F. Skinner, studied the way rats reacted to punishment and reward for performing certain tasks. Both men helped develop and reinforce the general scope of behaviorism: that the mind is capable of associating certain stimuli with certain observable behaviors or reactions (Cobb, 2008).

These theorists dominated the behaviorist school of thought for several decades. The problem, however, was that men like Pavlov and Skinner left no room for the possibility that the mind produces unobservable reactions to stimuli, too. Theorists like Jean Piaget believed that our minds are constantly trying to understand and reconcile the discrepancy between what we think we know about the world around us and what we actually find to be true. Piaget's theory of cognitive development introduced the concept of "schemas," which are topical collections of knowledge that include an array of both facts and interpretations drawn together in our minds based on what we have experienced. They are malleable and therefore allow us to assimilate and accommodate to new situations as they are presented to us in life (Halpenny & Pettersen, 2014).

Eventually, researchers recognized that behaviorism alone did not tell the whole story. If we can shape the way we react to certain stimuli just by observing the cause and effect relationship happening in the world around us, then we need to look at behaviorism and cognitive theory together. A movement brewed among behaviorists to marry the ideas of the two schools of thought, and the marriage eventually allowed for the distinction between overt and covert behaviors—the former being observable actions and the latter being internal actions like thoughts, feelings and perceptions (Cobb, 2008). It is through this combined perspective that we can better understand the process of anxiety development in the mind.

Under behaviorist theory, we can conclude that a person's fears are learned, and are the result of a cause and effect relationship in his or her mind. For example, a child who was badly bitten by a dog may become startled at the sight or sound of a dog later in life. It is reasonable to expect that he or she developed a conditioned response to a stimulus that frightened him or her. But what we cannot forget is that our minds are constantly processing new information from our environment. Under cognitive theory, we might see that the sibling of the child who was badly bitten by a dog developed a fear of dogs, too. The marriage of these two theories means that we can reasonably conclude that a person's fears can be learned as a result of an internal cause and effect relationship, an external cause and effect relationship that the person witnessed, or some combination of both.

Exploring anxiety through the lens of cognitive behavioral theory puts the client in the driver's seat of the therapeutic process and encourages his or her right to self-determination. The therapist, of course, must help the client navigate his or her mind and unearth the history of his or her own cognitive development, but it is up to the client to determine the root cause of his or her anxiety. As previously discussed, the DSM-5 puts the clinician in charge of telling the client what is or is not worth fearing. It might seem like it is obvious that some fears are blown out of proportion because they do not pose a physical threat (i.e. public speaking); but if we use that logic, where do we draw the line? The clinician cannot read the client's mind and cannot possibly understand the extent to which certain fears were conditioned in the client. If the clinician is quick to label a fear as irrational, this can lead to a misdiagnosis and a lack of proper treatment.

We previously discussed the ambiguous nature of the definition of anxiety in the DSM, and one other noticeable flaw is the absence of physical symptoms in the diagnostic criteria for most disorders outlined in this chapter. The only disorder that lists physical symptoms is panic disorder, and the American Psychiatric Association cites palpitations, racing heart, sweating, trembling, shortness of breath, chest pain, nausea and dizziness among the various psychosomatic symptoms that can be present (American Psychiatric Association, 2013). These are the paramount identifiers of a panic attack; though, some of these symptoms may be present in a client suffering from any one of the anxiety disorders listed in this chapter. This further complicates matters for a client and clinician trying to identify a diagnosis. In fact, some experts agree that clients are likely to suffer from more than one symptom—and possibly more than one disorder—and, therefore, we may be better off looking at the group of anxiety disorders as a whole rather than slicing up the disorders and symptoms into rather pointless categories (Gabbard, 2005).

If we looked at anxiety disorders as a whole, without separating them into discrete subcategories, we would have a better framework for understanding the cause and effect relationship of the mind. When clinicians are focused solely on trying to diagnose a certain anxiety disorder based on the impact the anxiety has on the client's life, they may be more likely to treat the symptoms instead of the cause. The DSM-5 makes a few basic assumptions about the client that are so overgeneralized that they border on unfair. As stated earlier in this paper, the American Psychiatric Association assumes that clients exacerbate their perceived fears, and thus render the client incapable of seeing the difference between rational and irrational fear. This line of thinking pigeonholes all clients into a category of psychosis where reality cannot be deciphered from irrational thinking.

However, most clients who suffer from an anxiety disorder know that their fears are irrational, but they cannot stop their mind from spinning out of control and into a state of panic when presented with that fear. Anxiety-producing fears—real or perceived—kick the adrenal gland into high gear, producing adrenaline and other stress hormones because our bodies are preparing us for fight or flight. The process has been in place as long as humans have, but the threats back then look very different from the threats we have now. Many of the anxiety-inducing situations clients face today are not situations where we must destroy our fear or flee from it, but our bodies still produce adrenaline that contributes to the onset of physical symptoms like racing heart, sweating, trembling, shortness of breath, etc. These symptoms are what impact the client's life the most, and under the guidelines of the DSM, clinicians are inclined to focus on eliminating those presenting problems. However, experts agree that while it is helpful to use breathing techniques, medicine, or other stress-reducers to alleviate the symptoms, these treatments should always be done in conjunction with psychotherapy (Gabbard, 2005).

In most cases, anxiety symptoms are the tip of the iceberg. While stress-reducing techniques can be beneficial in helping a client cope with severe symptoms and maintain some quality of life, they are not a substitute for discovering the client's underlying fear (Gabbard, 2005). Cognitive behavioral theory tells us that covert behaviors are learned just like overt behaviors, and a covert behavior that adapts over time into a cognitive distortion may take a lot of work to unlearn (Cobb, 2008). The DSM would lead us to believe that eradicating whatever symptoms the client is experiencing is the only thing that matters, but treating the symptoms alone will result in a short-term fix. Examining the situation from a cognitive behavioral perspective makes it obvious that the root of the problem is the most important factor in the therapeutic process. For any chance at lifelong results, clients must be talked through the process of cognitive restructuring to alter their problematic thoughts and behaviors.

However, taking back control of distorted thoughts is not easy. Clients have to face a lifetime of cognitive distortions, which can be incredibly challenging, overwhelming, and extremely emotional. However, much research is being done to improve treatment methods. According to a study published in the *Journal of Anxiety Disorders*, researchers are studying techniques to increase the success rate of anxiety treatments by mentally preparing the client for the road ahead at the onset of treatment. The researchers worked with patients who were diagnosed with generalized anxiety disorder and conducted motivational interviews prior to administering traditional cognitive behavioral therapy. The goal of the motivational interviewing was to reinforce the idea that change can be positive and good, despite clients' typical ambivalence to change. The researchers found that the technique holds promise for reducing anxiety symptoms during treatment (Westra, Arkowitz, & Dozois, 2009).

As clinicians, it is important to remember that clients who are in a heightened state of anxiety are essentially being controlled by their thoughts (Cobb, 2008). To make matters worse, anxiety is cyclical in nature. The more we allow ourselves to focus on the anxiety, the worse the physical and emotional distress gets. In 2007, researchers at the California Institute of Technology studied how this self-perpetuating worry can impact anxiety even before it starts. They aimed to answer the question of whether or not anxiety sensitivity—or, excessive worry about worrying—actually impacts our ability to cope during a bout of panic (Westra, Arkowitz, & Dozois, 2009). Some clients can fall so deep into bouts of worry about their next epi-sode of anxiety—i.e., their next panic attack, their next encounter with their phobia, where they will be, how they will handle it—that it debilitates their entire life. This study of 248 young adults analyzed how people have negative thoughts before, during and after experi-ences of anxiety, by means of a series of questionnaires. They found clinically significant results to conclude that anxiety sensitivity leads to excessive anxiety symptoms. The authors conclude that the more we worry about worrying, the worse the anxiety will be when it hap-pens next, and a key area of future research is in the way that we in-ternally regulate our feelings and our responses to feelings (Kashdan, Zvolensky, & McLeish, 2008).

The results of the anxiety sensitivity study reinforce the cognitive behavioral perspective that the cause of the anxiety really is im-portant. The study highlights the need for the client to be in tune with the emotions they are feeling when they are anticipating, experienc-ing, or reflecting back on a bout of anxiety. This forces the client to be mindful and aware of the cognitive distortions that are at play when anxiety comes about. Taking control of the cognitions and re-shaping them to be positive and productive can regulate the severity of the anxiety. Our thoughts control our anxiety, but we can take back control of our thoughts.

It is important to recognize that a healthy level of anxiety is actually quite normal, experts say, and if it is viewed strictly as a problem that must be eliminated, we are robbing clients of a launching pad for constructive self-doubt and problem solving (Gabbard, 2005). When the anxiety gets out of control, however, then we must help the client regulate the cognitions that are causing distress. According to a 2013 article published in *Frontiers in Human Neuroscience* there are two kinds of anxiety: adaptive and maladaptive. While adaptive anxiety is helpful for developing awareness and caution in dangerous situations, maladaptive anxiety can impact a person's cognition, force extra attention on negative events, and cause personal, social, and professional problems (Robinson, Vytal, Cornwell, & Grillon, 2013)

The anxiety disorders chapter of the DSM focuses on the client's irrational and exaggerated reactions to fears and aims at eliminating the subsequent anxiety symptoms, without any regard for the fact that the fear itself might be rational. For example, someone with arachnophobia has every reason and right to be afraid of a poisonous spider because an encounter with such a creature could cause fatal damage. Under the cognitive behavioral perspective we know that the fear itself is perfectly rational, but the cognitive distortions that have rooted themselves in the client's mind—causing anxiety when the person merely sees a photo of a spider or hears the word "spider"—are what need to be eliminated. We would be doing a great disservice to our clients if we aimed to completely eliminate their anxiety toward certain situations. But the DSM puts the clinician in charge of determining how irrational a fear is, and essentially assumes that the clinician knows more about the client's unique history than the client does. This simply cannot be true.

The reasons for the client's fear and anxiety are critical, yet they are not a priority in the DSM—nor is the client's say in the scale and scope of their own anxiety. Without first understanding what the root cause of the anxiety is, we cannot accurately diagnose a client with an anxiety disorder under the guidelines currently in the DSM-5.

References

Westra, H. A., Arkowitz, H., & Dozois, D. J. (2009). Adding a motivational interviewing pretreatment to cognitive behavioral therapy for generalized anxiety disorder: A preliminary randomized controlled trial. Journal of Anxiety Disorders, 23 (8), 1106–1117.

American Psychiatric Association. (2013). Diagnostic and Statistical Manual of Mental Disorders (Vol. 5). Arlington, VA: American Psychiatric Association.

Cobb, N. H. (2008). Cognitive-Behavioral Theory and Treatment. In N. Coady, & P. Lehmann (Eds.), Theoretical Perspectives for Direct Social Work Practice: A Generalist-Eclectic Approach (Vol. 2). New York, NY: Springer Publishing Company LLC.

Gabbard, G. O. (2005). Psychodynamic Psychiatry in Clinical Practice (Vol. 4). Arlington, VA: American Psychiatric Publishing Inc.

Halpenny, A., & Pettersen, J. (2014). Introducing Piaget: A guide for practitioners and students in early years education. New York, NY: Routledge.

Kashdan, T. B., Zvolensky, M. J., & McLeish, A. C. (2008). Anxiety Sensitivity and Affect Regulatory Strategies: Individual and Interactive Risk Factors for Anxiety-Related Symptoms. Journal of Anxiety Disorders, 22 (3), 429-440.

National Institute of Mental Health. (n.d.). Anxiety Disorders. Retrieved March 3, 2014, from National Institute of Mental Health: http://www.nimh.nih.gov/health/topics/anxiety-disorders/index.shtml

Robinson, O. J., Vytal, K., Cornwell, B. R., & Grillon, C. (2013, May). The impact of anxiety upon cognition: perspectives from human threat of shock studies. Frontiers in Human Neuroscience.

Young, Black, & Violent:
The Miseducation of Chief Keef

Whitney Pryor

Posttraumatic Stress in African American Males Living In Urban Environments

The conversation persists on how to solve the problem of inner-city violence in crime ridden, poverty stricken communities throughout the United States. It can be argued that no urban city, in modern times, has been plagued more with violent crime than Chicago. With more Chicago residents, 509 according to the Chicago Tribune, being killed in 2008 than American soldiers killed in the war in Iraq, 314 according to the Associated Press, the climate of violence and the presence of death in Chicago communities, that are populated by majority African American residents, lack the safety and security needed to make its residents feel that they are living in an environment that is conducive to producing happy, healthy, full lives.

Living in constant fear, and losing friends and family members to violence can leave anyone, especially young people, in a neverending state of trauma. Posttraumatic Stress Disorder, or PTSD, is detailed in the DSM-5 as follows:

Diagnostic Features: The essential feature of posttraumatic stress disorder (PTSD) is the development of characteristic symptoms following exposure to one or more traumatic events.

Diagnostic Criteria: The following criteria apply to adults, adolescents, and children older than age 6 years.

1. Exposure to actual or threatened death, serious injury, or sexual violence in one (or more) of the following ways:
2. Directly experiencing the traumatic event(s).

3. Witnessing, in person, the events(s) as it occurred to others.
4. Learning that the traumatic event(s) occurred to a close family member or close friend. In cases of actual or threatened death of a family member or friend, the event(s) must have been violent or accidental.
5. Experiencing repeated or extreme exposure to aversive details of the traumatic event(s) (e.g. first responders collecting human remains; police officers repeatedly exposed to details of child abuse).

Being exposed to violence on a daily basis leads young Black men to internalize feelings of fear and hopelessness. Most residents who live in these communities, where crime is prevalent, have to live there because they have no means of affording to live anywhere else. Often, these residents are uneducated, work low-wage jobs, and have multiple children to care for with very little support outside of these communities. Not only is violence a threat to daily survival, but it robs young people of the opportunity to enjoy time outside, to form relationships with others, and to focus on their futures, educationally, because they are in constant fear of death or injury. "Because trauma may interrupt the development of biological, cognitive, social and emotional knowledge, skills, and other resources, it can have a cascading and cumulative effect on a youth's life." (Stalker, 2008). The problem of violence and PTSD is then seen to have an effect, not just on the present life, of the young Black men living in violence, but it can also rob them of their futures. Often, unhealthy coping mechanisms are employed by young Black men such as drug use to numb the pain and fear of living with violence, or the carry of illegal weapons to protect themselves from being victimized. Both the former and latter have consequences that lend themselves to another epidemic that has spawn from the culture of violence in urban communities: the overrepresentation of young Black males in the prison population.

In Michelle Alexander's *The New Jim Crow* she explores the intersectionality of race, class and gender as it relates to the overwhelming number of African American men who are spending, as Erikson would deem it, their young adult era, ages 22-34 behind bars. A reality of living in violent spaces is that it is more comparable to a warzone than a community and it is often policed that way. In areas that are crime, drug, and gang infested the relationship that those who live in the community have with those who police the community is often not favorable. Not only are young Black males experiencing loss of life and safety, they are oftentimes experiencing the loss of their freedom as they react to and participate in violence and crime to survive in their neighborhoods. Alexander writes, "So far, we have seen that the legal rules governing the drug war ensure that extraordinary numbers of people will be swept into the criminal justice system—arrested on drug charges often for minor offenses...Once arrested, one's chances of ever being truly free of the system of control are slim, often to the vanishing point. Defendants are typically denied meaningful legal representation, pressured by the threat of a lengthy sentence into a plea bargain, and then placed under formal control—in prison or jail, on probation or parole." (Alexander, 2010). Therefore, a system, law enforcement, put in place to protect people and serve the community, injects an additional layer of trauma to those living in substandard conditions, in that, they have to not only face fear of death by the hands of their neighbors, but also fear of incarceration by a judicial system that is designed for them to be imprisoned.

As Alexander offered, the sentence of a person who is incarcerated for a drug crime does not end when their prison term ends. People who are released from prison with Class 1 or Class X drug felonies have to check a box for as long as that felony remains on their record that they have been convicted of a felony, which often means that they are unable to obtain a variety of jobs. It also means that they are ineligible for food stamps, public housing, federal student loan aid, and many are unable to exercise their right to vote. According to The Sentencing Project, a non-for-profit organization dedicated to research, advocacy, and reform, there are an estimated 5.85 million Americans who are unable to vote due to laws that prohibit voting by people with felony convictions. Included in that population, one in every thirteen African Americans is unable to vote due to felony drug convictions. Those who are freed from incarceration, but live in a systematic prison due to disenfranchisement, reenter society with limited means and access. They then contribute to the destitute nature of living in violence. Often, the experience of being imprisoned, only to return to an environment consumed with violence, crime and poverty, will lead them to live in a circular pattern of violence, crime, prison and trauma. "This disturbing phenomenon of people cycling in and out of prison, trapped by their second-class status, has been described by Loic Wacquant as a "closed circuit of perpetual marginality." Hundreds of thousands of people are released from prison every single year, only to find themselves locked out of the mainstream society and economy. Most ultimately return to prison, sometimes for the rest of their lives. Others are released again only to find themselves in precisely the circumstances they occupied before, unable to cope with the stigma of the prison label and the permanent pariah status." (Alexander, 2010). Meanwhile, they have children who have to live with the circular patterns of despair and enter that same life and/or live with PTSD as a result of their familial connection to, or individual practice of violent and unstable behavior.

Youth that have live in a constant fear of death or permanent injury and have no outlet to express that fear, often use music as a source of "self medication". Rap, or hip hop music, has provided an avenue for the young Black male to express his anguish and grief at the lost of his friends and family through street violence and or long term prison sentences. One of the most celebrated of all gangsta rappers, in Chicago, is teenager, Chief Keef. With rap lyrics that glorify violence, drug, use and casual sex, Chief Keef offers an insider's perspective to a day in the life of a teenager living in one of Chicago's most dangerous communities, Englewood. Chief Keef is 18 years old and cannot live in the city of Chicago because of fear that he will be shot or killed. He is currently out on parole for shooting at the police and already has a child and possibly has two women currently pregnant with his children. He is a high school dropout with gang affiliations and brandishes assault weapons on social media to tens of thousands of followers who desire to emulate the millionaire rapper. In the Chicago rap world, he is considered a God. His presence, along with other rappers who have left the "hood" for fame and fortune are often looked at as models of success and have set the standard for those in the "hood" to obtain. The lack of tangible, positive, role models leaves young, poor, urban Black men to aspire to live the type of life that ultimately leads to death, not super stardom. Recently, Chief Keef's cousin, a local rapper with the rap name Blood Money, who had just signed a major label record deal, was gunned down in Englewood, his dead body found on the sidewalk. Blood money is a reminder of those who try to live the life that Chief Keef lives, and it ends tragically as they fall victim to the life that they glorify in their music. The day after his cousin's death, Chief Keef is filmed entering court on a parole violation. He smiles and laughs as he enters the courtroom as though unfazed by the death of his cousin that happened hours prior. He, in essence, is the face of the young Black male living in violence, surrounded by drugs, death and prison time. He has become desensitized to death and unafraid of jail. He has learned to live with trauma and it has become a way of life. He is what the Black male with PTSD looks like. He is a problem, which America has yet to solve and until it does, more Black bodies will be in caskets or behind bars.

The issue of living in poverty and violence, with a kill or be killed mentality, not valuing life, has as much to do with the environment—physical—as it has to do with those who reside in those environments. Looking outside of, yet keeping in mind, the physical location, and considering the person-to-person relationship is important when continuing the conversation around posttraumatic stress disorder with those who live in violent environments. Considering the estimate from The Sentencing Project, in the chart below, the lifetime likelihood of imprisonment for African American men is 1 in 3 and is 1 in 18 for African American women. The likelihood that an adolescent Black male living in an urban environment will have a parent or relative who has had a prison experience is overwhelming high. The detriment that that has on an adolescent who is already suffering from poverty, substandard hosing, and less than desirable public schools, leaves the young Black male vulnerable to trauma as PTSD becomes a way of life. Being surrounded in a world that is unstable, young African American males are unable to build healthy, sustaining relationships and unable to have secure attachment to a parent or authority figure that is able to provide them with safety and shelter them from the burden of constant loss. "Attachment theory holds that psychological problems derive from the disturbances, deprivations, or disruptions in early caregiving relationships and from resulting distortions or limitations in internal representations of self, others, and relationships." (Stalker, 2008).

Attachment as a Behavioral System- "Attachment behavior is defined as "any form of behavior that results in a person attaining or maintaining proximity to some other clearly identified individual who is conceived as better able to cope with the world. (Bowlby, 1988). It is most apparent during childhood, but it can be observed throughout the life cycle, and especially when an individual, whether young or old, is frightened, injured, distressed, fatigued or ill. (Stalker, 2008). When a young person, or even an adult for that matter, is not able to securely attach to a healthy relationship, he or she then seeks to obtain that security in other ways, as including joining gangs and carrying handguns. Although these options lead to outcomes including death or imprisonment, the need to attach to others or a source of security outweighs the risk. "Bowlby (1969) emphasized the survival value of attachment in its provision of protection and safety, and he viewed this as the biological function of attachment behavior." (Stalker, 2008). Attachment theory offers that it is a person's biology to want to be safe and secure and that security is formed in infancy. When that security is lacking, the need to gain that security will be there until it is fulfilled. Although Bowlby does offer that an adult can gain secure attachment, the DMS-V offers that when a person has experienced more than one traumatic event they can suffer from "recurrent, involuntary, and intrusive distressing memories of the traumatic events." It can be offered that Black men living in constant trauma live with the consequences of that trauma, regardless of gaining positive secure attachment as they grow.

The lack of secure attachment or negative secure attachment can render a young Black man unable to deal with the issues of trauma and PTSD. While The DMS-V gives a multitude of examples of how one can be diagnosed with PTSD, and how different traumatic events lead to PTSD, it leaves much to be desired when speaking to the environmental factors leading to PSTD. For example, the idea of imprisonment is not included as an environmental factor, even though the imprisonment of young African American men is an epidemic in the United States. The DSM-5 also does not account for the trauma of chronic absentee fathers. Although being a child of a single parent is not inherently a traumatic experience, the overwhelming number of African American children growing up in broken homes or having to share a father with multiple family units as many African American households include a father who has multiple children with multiple mothers, the ability to gain trust and attachment with a male figure is lacking and can be a traumatic event as it an ongoing life stressor. The DSM-5 acts as a guide to explore what conditions a person has when they act outside of, or experience, actions that are deemed "normal". However, when considering cultures, what is seen as abnormal in majority culture can be considered normal in minority culture. As a minority, using the DSM-5 and applying PTSD characteristics to situations that are viewed as normal in minority culture can be disheartening, in that, as a minority, one is always striving to be "normal" or "similar" to majority culture, but yet one finds that minority normality does not fit into the scope of normal majority behavior. The DSM-5 does further strengthen the argument that normality is relative. Conversely, not finding incarceration or absentee parents in the category of environmental stressors offers that these things are not normal and should not be accepted as such, even if they are prevalent in African American communities.

Although mass incarceration and absentee fathers are issues that plague the African American community, there are initiatives that have been made throughout the country to curb the amount of violence happening in and around urban environments. The B.A.M. program, short for Becoming A Man, is based in Chicago, and works with minority male youth, in a mentoring environment to help with the absence of positive male role models in the lives of young boys. This local program, along with special schools like Urban Prep, in Chicago, which are set in urban areas like Englewood and only educate male students and boast a record of sending 100% of its graduating seniors to 4 year colleges, and national efforts like President Obama's My Brother's Keeper program work in concert to be an alternate, positive solution, for young minority males while allowing them to gain secure attachment to those who mentor and educate them. While these programs cannot stop the violence that happens in urban environments, they are a step in the right direction so that generations to come do not have to live in urban war zones and experience trauma as a normal way of life.

References

Alexander, M. (2010). *The New Jim Crow.* New York, NY: The New Press.

American Pyschiatric Assosciation. (2013). Diagnostic and statistical manual of mental disorders (5th ed). Washington, DC: American Psychiatric Association.

Digitale, E. (June 8, 2011).

Bonczar, T. (2003). Prevelance of Imprisonment in the US Population, 1974-2001. Washington DC: Bureau of Justice Statistics.

Childhood trauma linked to higher rates of mental health problems and obesity says Standford/Packard psychiatrist. *Stanford School of Medicine.* Retrieved from http://med.stanford.edu/ism/2011/june/carrion.html

Riddle, I. (August 11, 2013). In war on the streets, the similar threat of PTSD. *Nextcity.org.* Retrieved from http://nextcity.org/daily/entry/in-war-and-on-city-streets-the-similar-threat-of-ptsd

Rozas, Angela. (January 02, 2009). Chicago has 59 homicides in 2008. Retrieved from http://articles.chicagotribune.com/2009-01-02/news/0901010175_1_police-supt-jody-weis-crime

Stalker, C.A., and Hazelton, R. (2008), Attachment Theory, pp. 149-178. In N. Co-ady, & P. Lehmann (eds.). *Theoretical perspectives for direct social work practice.* 2nd. Ed, New York, N: Springer.

US Department of Veterans Affairs (January 3, 2014). Community Violence: The Effects of Children and Teens. Retrieved from http://www.ptsd.va.gov/public/types/violence/effects-community-violence-children.asp

The Micro Level Factor of Anorexia Nervosa

Robin Rogers

Many females in today's society believe that in order to be beautiful, they must be thin. For most females, this outlook leads to dieting and working out, but for others, it can lead to developing a feeding and eating disorder. "Feeding and eating disorders are characterized by a persistent disturbance of eating or eating-related behavior that results in the altered consumption or absorption of food and that significantly impairs physical health or psychosocial functioning," (Diagnostic and Statistical Manual of Mental Disorders [DSM-5], 2013). One eating disorder that society concentrates a lot of time on is anorexia nervosa (AN). Many say that young females develop anorexia because of the environmental and social factors, such as the media portraying skinny celebrities as beautiful. If these were the only main factors contributing to anorexia, then individuals with a feminist perspective would have great insight in diagnosing those with this disorder. Since these are not the only main factors, using the feminist perspective when diagnosing anorexia could potentially be harmful to a young female client. The one factor of anorexia that society and feminists tend to overlook is the micro level contribution: the biological factor. By adding this third component, it ensures that the individual's diagnosis is fully understood, rather than only partly understood. It also ensures that the client is seen as an individual rather than as another female in society affected by this disorder. Anorexia, and eating disorders in general, have gone through many phases, but "current theories have reached a tentative consensus that eating disorders rest squarely on the border between psyche and soma, fitting a truly biopsychosocial model" (Trace, Baker, Peñas-Lledó, & Bulik, 2013).

Anorexia Nervosa

According to the DSM-5, "there are three essential features of anorexia nervosa: persistent energy intake restriction; intense fear of gaining weight or of becoming fat, or persistent behavior that interferes with weight gain; and a disturbance in self-perceived weight or shape" (2013). The onset is in early adolescence and tends to affect females more than males, as the female-to-male ratio is 10:1 (DSM-5, 2013; Kaye, Wierenga, Bailer, Simmons, & Bischoff-Grethe, 2013). Although anorexia is one of the more discussed eating disorders within society, it is actually the least common, as "the prevalence is less than 1% of women" (Kaye, Wierenga, Bailer, Simmons, & Bischoff-Grethe, 2013). Even though it is the least common of the eating disorders, with "no proven treatments, AN has high rates of relapse, chronicity, and death" (Kaye, Wierenga, Bailer, Simmons, & Bischoff-Grethe, 2013). The starvation and extreme emaciation can also result in life-threatening medical conditions, such as effects on major organ systems and loss of bone mineral density (DSM-5, 2013). "Individuals with AN tend to have other puzzling symptoms and behaviors that are poorly understood, such as severe body image distortions, a lack of insight about being ill, and depriving themselves of food despite starvation" (Kaye, Wierenga, Bailer, Simmons, & Bischoff-Grethe, 2013). "Anorexia brings with it the appearance and feeling of total control and almost total denial" (Burns, 2004).

Anorexia nervosa occurs all over the world. In 1996, Sing Lee stated that, "Recent attention, however, has been drawn by researchers to the emergence of the cultural fear of fatness and AN not only in different socioeconomic levels and minority groups in Western societies, but also from diverse countries in Europe, Africa, India, The Middle East and The Far East." Since each country is so different and anorexia might be "bound to the culture of modernity," the "evidence suggests cross-cultural variation in its occurrence and presentation"

(DSM-5, 2013; Lee, 1996). In regards to the United States, it should be noted that "anorexia nervosa appears comparatively low among Latinos, African Americans, and Asians" because these groups are less likely to ask for help or have the resources to receive help (DSM-5, 2013). Overall, "individuals with anorexia nervosa may exhibit a range of functional limitations" such as, "social isolation and/or failure to fulfill academic or career potential" (DSM-5, 2013). "The crude mortality rate for anorexia nervosa is approximately 5% per decade" and "death most commonly results from medical complications associated wit the disorder itself or from suicide" (DSM-5, 2013).

Feminist Theory

"Feminist theory is directed toward bringing about a future better than and different from the present" (Grosz, 2010). There are many feminist theories, and they each tend to take their own approach in reaching this universal goal of change. Each theory has its own history and its own target audience. There is lesbian feminism, which is "critical of heterosexual institutions," and womanism, which "starts with the perspective of black women and centers on a complex matrix of oppressions" (Coady and Lehmann, 2008). Another example is radical feminism, which argues, "that individual women's experiences of injustice and the miseries that women think of as personal problems are actually political issues, grounded in power imbalances" (Coady and Lehmann, 2008).

In general, feminist theories analyze "the varied experiences of people from a political perspective" and these experiences "can best be understood in terms of sex-based and gender-based social and structural restrictions, constrictions, and resource deficits" (Coady & Lehmann, 2008). "Feminist epistemology (or feminist theory more generally) often presents itself as a vehicle through which women can come to find their own voices" (Crary, 2001). Also, "the utility of psychotherapy is hotly debated" as, "some would argue that the goals of psychotherapy – individual change in behavior and self-perception – distract feminist social workers, community members, and service users from social justice work" (Coady & Lehmann, 2008). By focusing on the injustices towards women, feminist theories usually practice at the macro level. They work to bring about changes within society and they do this by enabling "women to empower themselves as individuals and collectively" (Coady & Lehmann, 2008).

Critique of Feminist Theory as a Treatment for Anorexia Nervosa

When diagnosing an individual with anorexia nervosa, it is important not to lose sight of who the person is that is suffering. Often times, the diagnosis will immediately be blamed on macro level influences, such as the media or the recent cultural trend of being thin. For instance, feminist theorists "have been pivotal in shifting interpretations of anorexia away from a strictly clinical domain and linking it to the impact of oppressive cultural systems, such as patriarchy and the mass-media" (Burke, 2006). By viewing anorexia in this way, feminist theorists only see the diagnosis, rather than the individual with the diagnosis. Also, from this perspective, they seem more concerned with changing society's oppressive systems, than treating the clients with the disorder. "The problematic oral schema of anorexia is linked through consumption to other forms of consumption not directly related to food, most notably consuming imagery and literature in which slenderness is presented as a desirable ideal" (Burke, 2006). Feminist theorists can spend time arguing that anorexia occurs because of the consumption of imagery and literature, but how can they explain the millions of other girls who also consume this information and never develop anorexia? There are girls every day reading pop-culture magazines, dealing with difficult family situations or trying to impress boys at school who develop healthy eating or exercising habits, instead of anorexia. By looking at anorexia as a political, gender related or environmental issue, feminist theorists risk overlooking other important contributions, such as the micro level factors.

Many feminist theorists use an assessment process that includes "examining individual circumstances, a person's membership in a particular socioeconomic class, ethnic group, and racial category", and "information about populations who have comparable strengths, experience similar challenges, and are affected by the same type of sociopolitical circumstances" (Coady & Lehmann, 2008). These steps are completed in order to determine if there should be a micro and/or macro level intervention (Coady & Lehmann, 2008). Since there is a choice to intervene on one or both of the levels, there is a possibility that feminist theorists overlook key aspects of the individual. Also, it seems as though the third step to this assessment can be very diminishing to the individual, as individuals diagnosed with anorexia can differ greatly from one another.

One differing factor is an individual's genetic predisposition to anorexia. "First-degree relatives of individuals with AN are 11 times more likely to have lifetime AN than are relatives of unaffected controls" (Trace, Baker, Peñas-Lledó, and Bulik, 2013). Also, as stated in the DSM-5, "Concordance rates for anorexia nervosa in monozygotic twins are significantly higher than those for dizygotic twins" (2013). Even if the client does not see the importance in family history, in this case, it makes sense to learn about the biological information. Many feminist theorists would either not find out or skip over this information, but in reality, this would be imperative information to know when assessing a client at the beginning of treatment. "Although genes clearly contribute to eating disorder liability, they do not act alone but rather in concert with environmental factors" (Trace, Baker, Peñas-Lledó, and Bulik, 2013). Without considering this genetic contribution, feminist theorists are at risk of only focusing on the macro piece and potentially missing key information.

This separation might also happen if feminist theorists neglect another potential micro level factor, behavioral/personality traits, which can predispose individuals to anorexia. "The traits include anxiety, negative emotionality, perfectionism, inflexibility, HA, and obsessive behaviors (particularly with order, exactness, and symmetry)" (Kaye, Wierenga, Bailer, Simmons, & Bischoff-Grethe, 2013). An additional trait connected to anorexia is "inhibitory self-control and reward" (Kaye, Wierenga, Bailer, Simmons, & Bischoff-Grethe, 2013). "Individuals with AN have long been noted to be anhedonic and ascetic, able to sustain the self-denial not only of food but also of most comforts and pleasures in life" (Kaye, Wierenga, Bailer, Simmons, & Bischoff-Grethe, 2013).

Lastly, another main behavioral factor that has been linked to anorexia is attention bias (AB). This "refers to a tendency to over focus awareness in the environment that is disorder salient" (Aspen, Darcy, & Lock, 2013). For those with anorexia, the focus would be on body weight and shape (Aspen, Darcy, & Lock, 2013). "It has been theorized that AB can predispose an individual to develop a disorder and/or operate to maintain it" (Aspen, Darcy, & Lock, 2013). Many individuals diagnosed with anorexia might not see these behavioral factors as a problem or they might not even recognize that they have these vulnerable traits. If this is the case, then many feminist theorists would not address them at the intervention phase because using a feminist intervention means giving "service users the specific help they ask for" (Coady & Lehmann, 2008). Since it is possible that these traits can lead to or maintain anorexia, then understanding an individual's personality traits is imperative for the treatment process. Recognizing a person's behavioral traits also shows that the theorist sees the whole individual, rather than just the person's issue with body weight. Overall, this feminist intervention plan could really fail an individual with anorexia, especially since "individuals with anorexia nervosa frequently either lack insight into or deny the problem" (DSM-5, 2013).

Conclusion

Since there are many contributing factors to anorexia nervosa, it is critical that all are addressed in order to provide the utmost help to the individual and limit the relapse rate. Although the macro level factors might play a bigger role in the development of anorexia, the micro level factors still exist and need to be addressed. For example, in a performance, the audience can only see the actor's hard work, but without the tech and costume crew, there could be no show. The genetic and behavioral factors contributing to anorexia might be "behind the scenes" but without them, the disorder might not have developed. Feminist theorists tend to dismiss these micro factors and place the emphasis of the disorder on the problems of the greater world. From this view, the diagnosis and the individual are lessened, and the emphasis is placed on how to stop anorexia in all girls, rather than this one girl. Also, by attributing the diagnosis of anorexia to a gender, political or macro-based issue, then feminist theorists are enabling the individual to blame society for his or her problems. This attribution takes the responsibility for recovery away from the individual and places it on everyone else. With this in mind, feminist theorists should not take this macro approach when working with individuals with anorexia because it could be detrimental to each individual's treatment process.

References

Aspen, V., Darcy, A., & Lock, J. (2013). A Review of Attention Biases in Women with Eating Disorders. *Cognition and Emotion*, *27*(5). Retrieved from http://www.tandfonline.com.flagship.luc.edu/doi/abs/10.1080/02699931.2012.749777#.U7tBpRbmpZ4

Burke, E. (2006). Feminine Visions: Anorexia and Contagion in Pop Discourse. *Feminist Media Studies, 6,* 316-329. doi: 10.1080/14680770600802066

Burns, M. (2004). Eating Like an Ox: Femininity and Dualistic Constructions of Bulimia and Anorexia. *Feminism and Psychology, 14,* 269-295. doi: 10.1177/0959-353504042182

Coady, N. & Lehmann, P. (2008). *Theoretical Perspectives for Direct Social Work Practice* (2nd ed.). New York, NY: Springer Publishing Company, LLC.

Crary, A. (2001). A Question of Silence: Feminist Theory and Women's Voices. *Philosophy,* 371-395. doi: 10.1017/S0031819101000432

Diagnostic and Statistical Manual of Mental Disorders. (2013). Arlington, VA: American Psychiatric Association.

Grosz, E. (2010). The Practice of Feminist Theory. *A Journal of Feminist Cultural Studies, 21,* 94-108. doi: 10.1215/10407391-2009-oig

Kaye, W. H., Wierenga, C. E., Bailer, U. F., Simmons, A. N., & Bischoff-Grethe, A. (2013). Nothing Tastes as Good as Skinny Feels: The Neurobiology of Anorexia Nervosa. *Trends in Neurosciences, 36.* Retrieved from http://www.sciencedirect.com.flagship.luc.edu/science/article/pii/S0166223613000064

Lee, S. (1996). Reconsidering the Status of Anorexia Nervosa as a Western Culture-Bound Syndrome. *Social Science & Medicine, 42.* Retrieved from http://ac.els-cdn.com/0277953695000747/1-s2.0-0277953695000747-main.pdf?_tid=ee03d4d8-19e8-11e4-b383-00000aacb35d&acdnat=1406945074_b8217b38f8454765c1c1d264409dd34a

Trace, S. E., Baker, J. H., Peñas-Lledó, E., & Bulik, C. M. (2013). The Genetics of Eating Disorders. *The Annual Review of Clinical Psychology, 9.* Retrieved from http://www.annualreviews.org.flagship.luc.edu/doi/pdf/10.1146/annurev-clinpsy-050212-185546

Distancing Mind/Body Dualism from Somatic Symptom Disorder Diagnosis

Amy Sandquist

Introduction

Somatic symptoms are physical symptoms that occur without bi-omedical explanation. Unexplained headaches, backaches, and stom-achaches are common examples of somatic pain, and many people are familiar with such mysterious but persistent pains. In fact, Rief and Martin (2014) cite a 2006 study by Hiller et al. estimating that "up to 80% of the general population reports somatic symptoms during the last seven days" (p. 340). Somatic symptoms are arguably a banal part of human life: Sometimes our bodies hurt, and we cannot explain the source of pain. When these symptoms become a hindrance to daily living, however, the DSM-5 categorizes them as pathology, as somatic symptom disorder (American Psychological Association, 2013, p. 309). Thus, unexplained physical pain becomes a psychological pathology when the somatic symptoms cause "significant stress and impairment" (American Psychological Association, 2013, p. 309). Naturally, chronic pain can inhibit people's well-being by preventing them from working, keeping them from socializing with friends and family, and decreasing their ability to be active.

Under the somatic symptom and related disorders category, the DSM-5 includes somatic symptom disorder, illness anxiety disorder, conversion disorder, and factitious disorder (American Psychological Association, 2013). Throughout this paper, I primarily will refer to and discuss somatic symptom disorder, a diagnosis that requires "one or more somatic symptoms that are distressing or result in significant disruption of daily life," "excessive thoughts, feelings, or behaviors related to the somatic symptoms," and persistent symptomatic experiences lasting at least six months (American Psychological Association, 2013, p. 311). The DSM-5 stresses that excessive feelings and thoughts, which can include "disproportionate and persistent thoughts," "persistently high [levels] of anxiety about health or symptoms," and "excessive time and energy devoted to these symptoms or health concerns," are the crucial criteria for diagnosis (American Psychological Association, 2013, p. 312). Thus, somatic symptom disorder is characterized by two types of pain: physical and psychological.

Not only do people suffering from somatic symptom disorder have to deal with chronic pain or discomfort, but sufferers also have to cope with social stigma about their pain. When medical professionals cannot find the cause of patients' chronic pain, the patients are often viewed as dishonest or attention seeking. Redekop, Stuart, and Mertens (1999) write that people are labeled with somatic symptoms with the implicit understanding that "their symptoms are under volitional control" and that these patients "use their physical symptoms as a way to manipulate and control others" (p. 1). Further, the authors claim, "'Somatizing and 'somatoform' have become thinly-veiled epithets directed at such patients," implying that many medical practitioners use the diagnoses out of frustration (p. 1).

The American Psychological Association is working to counteract such stigma by decreasing the extent to which mind/body dualism—the idea that the mind and the body are two distinct entities—comes into play in the diagnosis (Rief & Martin, 2014). Rief and Martin (2014) suggest that the DSM-5's characterization of somatic symptom disorder, instead of the somatoform disorder that Redekop, Stuart, and Mertens (1999) describe from the DSM-IV, attempted to decrease the stigma associated with the physical symptoms. One of the most significant changes between the DSM-IV and DSM-5 was the diagnostic category's name: The DSM-IV diagnostic category was called somatoform disorders, and, as already discussed, the DSM-5 diagnostic category is called somatic symptom and related disorders (Rief and Martin, 2014). Rief and Martin (2014) suggest that this change discourages mind/body dualism, as it emphasizes the physical symptoms instead of suggesting that "everything is just in the mind" (Rief and Martin, 2014, p. 341). Furthermore, the DSM-5 "[abolished] the distinction between medically explained and medically unexplained complaints," which means that it does not matter whether or not the sufferer is experiencing pain that is medically explainable (Rief and Martin, 2014, p. 341). These changes pave the way for a more integrated discussion of the relationship between the psychological and the physical, between the mind and body.

Contemplating somatic symptom disorder gives social work practitioners the opportunity to explore the relationship between people's minds and their bodies. Characterized by the DSM-5 as a treatable psychological disorder, somatic symptom disorder causes patients to experience real physical pain. Often, people presenting somatic symptoms seek help from non-psychiatric physicians before they are referred to psychologists or psychiatrists (American Psychological Association, 2013, p. 309). This complex relationship between psychological distress and physical pain could point to an uncomfortable truth about Western culture (the culture for which the DSM-5 was written): Expressing physical pain is a more acceptable and less stigmatized way than expressing psychological distress to receive treatment and relief. Furthermore, treatments for physical pain could seem more direct, more efficacious, and more painless than treatments for psychological distress. Taking a prescription for pain medication, for example, is a more passive and, quite frankly, an easier way to obtain pain relief than engaging in long-term psychological treatment. Therefore, expressing pain through physical symptoms might be a less stigmatizing method than expressing pain psychologically.

In order to examine people's understanding of the relationship between physical and psychological pain, we must first explore ways in which ideas about the physical and the psychological have been constructed in U.S. society. Social construction theory can help practitioners understand that commonly accepted notions about reality are constructed based on ideology. Social construction theory posits, "'Society is a human product. Society is an objective reality. Man is a social product'" (Berger & Luckmann as qtd. in Robbins, Chatterjee, & Canda, 2012, p. 329). Thus, social constructionists argue that reality is culturally molded and pre-defined (Robbins, Chatterjee, & Canda, 2012). In other words, different cultures construct their own theories of reality based upon the values that cultures deem worthy. To define the relationship between the psychological and the physical, the West, for instance, tends to ascribe to a view first written about by Rene Descartes in the sixteenth century, a view that explains human life as the interaction between "the 'errors' of the body" and "the 'truths' of the mind" (Redekop, Stuart, & Mertens, 1999, p. 7). This Cartesian approach, creating a distinct duality between the mind and the body, has "influenced the Western approach to economic, political, legal, cultural, and administrative systems" (Jun, 2005, p. 88). In other words, the mind/body dualism enacted by Descartes has permeated Western life and U.S. culture.

While mind/body dualism might not seem harmful in itself, the idea can perpetuate the objectification of bodies and isolate both psychological and physical pain from their larger environmental causes. To combat these consequences, the DSM-5 is already working toward integrating practitioners' understanding of the false dichotomy between the mind and the body (Rief & Martin, 2014). In order for the dichotomy to be truly abolished, however, practitioners in the United States must understand the extent to which U.S. society promotes an understanding of the mind and body as two separate entities. Using social constructivist theory and feminist theory, we can deconstruct the Western notion that body and mind are separate and move toward a more integrated understanding of mind/body unity. With a clearer understanding of mind/body unity, social work practitioners can begin to examine the ways in which somatic symptom disorders can provide sufferers with concrete ways of expressing their psychic discomfort and of reasserting a degree of control over their bodies.

Constructivism and Mind/Body Dualism

For centuries, people have debated the relationship between the human mind and the human body. The belief in mind/body dualism can be found around the world and in many different cultures (Forstmann, Burgmer, & Mussweiler, 2012). Some scholars argue that mind/body dualism facilitates basic spiritual beliefs, such as "beliefs in heaven and hell, reincarnation, and spirits of deceased family members" (Bering, 2006; Bloom, 2007 as cited in Forstmann, Burgmer, & Mussweiler, 2012, p. 1239). Forstmann, Burgmer, and Mussweiler (2012) present dualism as an innate human trait, "an evolutionary accident," and quote Bloom (2007), who views dualistic beliefs as "a natural by-product of the fact that humans have two distinct cognitive systems, one for dealing with material objects, the other for social entities" (p. 149 as qtd. in Forstmann, Burgmer, & Mussweiler, 2012, p. 1240). Despite the fact that some scholars believe mind/body duality to be an innate human trait, some cultures emphasize mind/body dualism more than others. Eastern traditions, like Confucianism, Buddhism, and Hinduism, for example, maintain belief systems that allow for the integration of the mind and body (Jun, 2005).

Numerous scholars acknowledge the influence that Descartes' mind/body dualism has had on Western life (Jun, 2005; Redekop, Stuart, & Mertens, 1999; Forstmann, Burgmer, & Mussweiler, 2012; Duncan, 2000). In his famous text *The Treatise of Man*, Descartes separated the body—a being's physicality and metabolic functions—from the soul, which he defined as a being's ability to think rationally (Duncan, 2000, p. 506). In fact, Descartes defined the body and soul as "totally different substances, the body being spatially extended, non-thinking, and divisible . . .; and the soul being thinking, indivisible, and non-extending" (Duncan, 2000, p. 488). Such a distinction defined humanity: Without the rational soul, Descartes believed that people would be no different than animals (Duncan, 2000, p. 506). Duncan (2000) pointed out that Cartesian dualism—Descartes' famous division of body and soul—acknowledges that the body and soul are closely aligned while the human being is alive. However, the dualism remains vital, as it allows Descartes to easily support Christian belief structures that maintain that the soul leaves the body upon death (Duncan, 2000). As such, Descartes' conceives that the body is the finite component that limits humanity, and the soul, or mind, is the infinite—and even Godly—force that lives on after the body's death. Thus, Cartesian dualism considers the mind to be the more powerful, humanizing force. This value-laden dualism set the stage for the West's emphasis on the mind's primacy and the body's limitations. Cartesian dualism advocates the view that the body cannot be trusted, an implication that still stigmatizes those experiencing somatic symptom disorders.

Deconstructing Mind/Body Dualism with Feminist Theory

As previously discussed, assigning value to each pole of the mind/body spectrum creates inequality between the physical and the psychological, positing the two in opposition with one another. Feminist theory can help practitioners understand how socially constructed mind/body dualism creates an unequal relationship between the mind and the body. In her text *Feminist Theory: From Margin to Center*, hooks (2000) argued that feminism must be concerned with "eradicating the politic of domination," meaning that feminism must broaden its goal of dismantling gender inequity to dismantling entire systems of oppression and domination, such as racism and capitalism (p. 21).

The West's socially constructed mind/body dualism enacts a system of oppression in which bodies are oppressed by the minds inside of them and, in a larger sense, oppressed by profit-seeking corporations that set standards for physical health and beauty. When the mind and body are considered separate entities, they inevitably become unequal. And, in the U.S., corporations profit by convincing consumers that the body is a distinct, inferior entity that needs to be controlled.

The West's emphasis on separating mind from body hurts both men and women, but the separation harms women more acutely. While objectifying women's bodies is not an inherently Western concept, Western culture, media, and politics do their best to isolate women's bodies as units that require constant evaluation and superficial care. In fact, in the U.S., women's bodies serve as physical reminders of social norms and values. For example, U.S. culture values youth, virility, and attractiveness to such an extent that U.S. society increasingly pathologizes women's aging bodies, as the bodies no longer represent the youthful ideal.

This pathologizing drastically transforms the ways in which women relate to their aging bodies, which society encourages women to view as liabilities. Brooks (2010) described the relationship between the availability of anti-aging products and surgery and women's increased responsibility for "fixing" their aging faces and bodies (p. 247). Brooks contended, "The very fact that women can make the choice to have and use [anti-aging technologies] . . . makes some women feel personally responsible for their aging faces and bodies. Ageing becomes *their fault*" (p. 247). Thus, women's bodies become capitalist commodities, items to be bought and sold, and as product developers and corporate marketers gain control over the ways that women's bodies should look, smell, taste, and feel, women slowly lose control over their bodies, and as they lose control over their bodies, women lose agency.

Not only are women's bodies brought into popular cultural discourse, but women's bodies are violently propelled into national debate. Debates about abortion—whether or not a woman has the right to terminate her own pregnancy—are emotion-laden, shaped by religious devotion, and highly volatile. However, importantly, the U.S.'s debates about abortion and birth control hit at the heart of the separation of mind and body: The debate questions whether women can use their intellectual capacities to control their own bodies. While the debate about abortion is contentious, the debate itself—the question of the legality and morality of abortion—implicitly tells women that their bodies' reproductive functions are separate from their intellectual functions and that they cannot make choices based on the integration of their minds and bodies.

Therefore, as U.S. medicine and consumerist culture objectify bodies, the cultural forces also rob people, particularly women, of a measure of control over their bodies. U.S. culture asserts a popular mind/body dualism that prevents people from managing the relationship between their bodies and minds. Interestingly, somatic symptoms disorder also presents as a loss of control over one's body: chronic, long lasting, and sometimes inexplicable pain, combined with rumination or anxiety about the pain. Unsurprisingly, perhaps, women are "more than six times as likely to be diagnosed with somatization disorder than men" (Swartz, Blazer, George, & Landerman, 1986 as cited in Redekop, Stuart, & Mertens, 1999). Only with the concept of mind/body dualism can we properly understand the ways in which somatic symptom disorder allows people to assert control over their bodies.

Bodies as Tools of Psychological Expression

By considering the mind and body as two parts of one whole, practitioners can better understand the pain experienced by people suffering from somatic symptom disorder. In bodies coded by U.S.

consumer culture and dominated by an oppressive patriarchal social structure, people might feel helpless and unable to maintain control over their lives. These feelings of helplessness might manifest as psychological distress, which, simultaneously, could manifest as uncontrollable physical symptoms. This section seeks to illustrate the ways in which physical pain can serve as an indication of psychological distress, causing the sufferer distress while at the same time allowing her concrete means for expression.

In order to illustrate practitioners' tendencies to focus on either the physical or the psychological instead of integrating the two, I will examine Katzmann and Lee's (1997) article about eating disorders. I am not arguing in favor of any similarities or differences between eating disorders and somatic symptom disorders; rather, I am exploring Katzmann and Lee's (1997) argument so that I can use their argument as a frame through which to better understand mind/body dualism and somatic symptom disorder. Katzman and Lee (1997) explain that conventional wisdom—and many medical professionals—suggest that eating disorders stem from low self-esteem and poor body image. The scholars see the focus on low self-esteem and body image as problematic, as it implies that eating disorder sufferers care solely about their physical presentation. As Katzman and Lee (1997) state, "By emphasizing slenderness, the dominant imagery about eating concerns misnames as much as it discounts real biases against women and their limited access to other forms of power of self-expression beyond corporeal power" (p. 389). Therefore, when physicians, psychologists, and social workers fail to acknowledge the deeper implications of physically manifested psychological distress, they inadvertently maintain oppressive constructs and fail to understand the real causes of the sufferers' pain. In order to fully understand somatic disorders, mental health professionals must also understand the social and political environment in which their clients live.

By focusing on clients' social, cultural, political, and economic environments, practitioners can glean a more unified and holistic perspective about the relationship between clients' physical and psychological distress. Rather than categorizing eating disorders as "fear of fatness" (p. 388), Katzman and Lee (1997) argued that eating disorders could be mechanisms for asserting control and agency. To further their point, Katzman and Lee (1997) described psychiatrist and anthropologist Littlewood's (1995) assertion that "South and East Asian women, in the absence of fear of fatness, engage in self-starvation to instrumentally achieve self-determination when confronted with ambivalent cultural demands" (p. 388). That is, some women engage in disordered eating not to achieve an aesthetic ideal but rather to practice an asceticism that allows them greater control over their lives. In other words, eating disorders could be interpreted as physical manifestations of psychological resistance to oppressive cultural norms.

Similarly, I contend that somatic symptom disorders can represent physical manifestations of psychological resistance to oppressive culturally sanctioned norms. When practitioners consider the body and mind to be separate entities, somatic symptom disorder appears to be a disorder in which the body is thwarting the mind. However, when practitioners understand the body and mind as one whole, they are able to look beyond the suffering individual's corporeal reality to the individual's broader social environment, and somatic symptom disorders can be understood as whole-mind/body assertions of power.

When mind/body dualism is abolished, practitioners can reach a deeper understanding about somatic symptom disorders. In her recent article, Marrow (2013) looked beyond the somatic symptoms of sufferers' physical presentations and potential psychological disturbances to the fraught and contradictory cultural, social, political, and economic expectations heaped onto the young women experiencing symptoms. Marrow (2013) examined "clenched teeth," a somatic phenomenon that occurs in North India when young women display a "sudden unresponsiveness, a fainting-like or seizure-like episode, and a tightly closed mouth" (p. 347). Marrow (2013) presented two case studies and argued that both cases illustrate examples of young women negotiating "the contradictory and fraught consequences of macrolevel social changes permeating the environments of non-elite middle-class girls and women" (p. 349).

In other words, presenting physical symptoms allowed these girls to distance themselves from the complex social dilemmas faced by many women around the world. In her fieldwork, Marrow (2013) noted that the girls she observed were enduring "expectations of upward mobility, education, and well-paying jobs as a result of globalization" while also facing an "increasingly rigid definition of gender and family roles" (p. 349). Therefore, on a perhaps unconscious level, the young women defy their complex and socially prescribed roles by manifesting physical symptoms, which do little to solve the dilemmas the young women face but which do allow them to have "crises of responsibility in the mother-daughter relationship," forcing their mothers to work with them to theorize the girls' futures amidst conflicting social and familial demands (Marrow, 2013, p. 357).

Marrow's argument clearly demonstrates how somatic symptom disorder can be a mechanism for expressing psychological distress with physical symptoms. The young women Marrow described do not live in the West, where mind/body dualism is a more popular philosophical stance, yet they still fall prey to contradictory expectations that society heaps onto them. In Marrow's conceptualization, the young women benefit from the "clenched teeth" episodes because the episodes provide space for the women and their mothers to "work through" questions about social expectations and female agency (Marrow, 2013). While Marrow's speculation about these women's gains is valuable, the gains she discusses are culture-specific. Overall, though, somatic symptom disorder sufferers do not necessarily gain anything concrete—they suffer from chronic symptoms and psychological distress. This paper contends that when the mind and the body are recognized as one unit, somatic symptom disorder sufferers gain the ability to express, but not articulate, resistance to socially constructed oppression.

Conclusions and Implications for Practice

The DSM-5's editors made a responsible decision as they sought to limit stigma against somatic symptom disordered patients by limiting the amount of mind/body duality in the somatic symptom disorder criteria. While mind/body duality is not inherently bad or wrong, limiting the extent to which practitioners rely on mind/body duality can help practitioners understand their clients as people who exist within particular biological, psychological, social, and political environments.

Identifying and assessing oppressive and socially constructed forces can help the mental health community resist stigmatizing somatic symptom sufferers. However, such macro-level analysis might have limited implications for micro practice. Redekop, Stuart, and Mertens (1999) warn practitioners that when they speculate that a patient's symptoms are caused by the "unjust structure of society," they risk sending a message that "may not be comforting in an immediate sense, since it may be interpreted as saying that [a patient's] illness is just as 'made up'" (p. 3). Thus, explanations for pain that revolve around social constructivism and feminist theory might be vital for sociologists, social workers, and medical professionals to consider, but such explanations can also sound like distractions from the real, and often chronic, pain felt by the sufferer. Practitioners should educate clients about mind/body unity by teaching mindfulness and meditation; however, practitioners must also respect clients' presenting

problems and treat clients' complaints of chronic pain with understanding and compassion.

References

American Psychological Association. (2013). *Diagnostic and statistical manual of mental disorders* (5th ed.). Arlington, VA: American Psychiatric Publishing.

Brooks, A.T. (2010). Aesthetic anti-ageing surgery and technology: Women's friend or foe? *Sociology of Health & Illness* (32) pp. 238-257 doi: 10.1111/j.1467-9566.2009.01224.x

Duncan, G. (2000). Mind-body dualism and the biopsychosocial model of pain: What did Descartes really say? *Journal of Medicine and Philosophy*, (25)4. Pp. 483-513.

Fisher, M.A. (2007). Medicine and industry: A necessary but conflicted relation-ship. *Perspectives in Biology and Medicine*, (50)1, pp. 1-6. Doi: 10.1353/pbm.2007.0003

Fortsmann, M., Burgmer, P., & Mussweiler, T. (2012). "The mind is willing, but the flesh is weak": The effects of mind-body dualism on health behavior. *Psychological Science*, (23). Pp. 1239-1245. Doi: 10.1177/0956797612442392

hooks, b. (2000). Feminist theory: From margin to center. South End Press Classics. Cambridge.

Jun, J.S. (2005). The self in the social construction of organizational reality: Eastern and Western views. *Administrative Theory & Praxis*, (27)1. Pp. 86-110.

Katzman, M.A. & Lee, S. (1997). Beyond body image: The integration of feminist and transcultural theories in the understanding of self starvation. International Journal of Eating Disorders, (22). Pp. 385-394.

Marrow, J. (2013). Feminine power or feminine weakness?: North Indian girls' struggles with aspiration, agency, and psychosomatic illness. *American Ethnologist* (40), pp. 347-351. Doi: 10.1111/amet.12026

Redekop, F., Stuart, S., & Mertens, C. (1999). Physical "phantasies and family functions: Overcoming the mind/body dualism in somatization. *Family Processes,* (38). Pp. 371-385.

Rief, W. & Martin, A. (2014). How to use the new DSM-5 somatic symptom disorder diagnosis in research and practice: A critical evaluation and a proposal for modifications. *Annual Review of Clinical Psychology*. Doi: 10.1146/annurev-clinpsy-032813-153745

Robbins, S.P., Chatterjee, P., & Canda, E.R. (2012). *Contemporary human behavior theory: A critical perspective for social work* (3rd ed.). Upper Saddle River, NJ: Pearson Education, Inc.

The Paradox of Psychiatry as a Medical Science and Product Of Social Constructivism

Allison Stein

History of the Diagnostic and Statistical Manual of Mental Disorders

Today's mental health professions originated in the early nineteenth century in response to the American government's need for classification of mental disorders. In that era, however, mental disorders as we know them today did not exist. What existed instead were behaviors that reflected pathology that were later organized into diagnostic categories of the Diagnostic and Statistical Manual of Mental Disorders (DSM). In 1840, the first unofficial DSM included two diagnostic categories: idiocy and insanity. Forty years later, the DSM expanded to seven categories including mania, melancholia, monomania, paresis, dementia, dipsomania and epilepsy. By 1942, a committee of the American Medico-Psychological Association (now the American Psychiatric Association) had made a total of ten revisions to the manual. Ten years later the first official edition of the DSM was published with a total of 106 "mental reactions," which inferred the "psychobiological view that mental disorders represented reactions of the personality to psychological, social, and biological factors" (DSM-IV-TR, 2000, p. 153).

Significant shifts in the US psychiatric classification system were largely due to the need to classify, and more importantly, clarify, the behaviors seen in state hospitals. Behaviors and symptoms of modern day clinical depression, schizophrenia, and bipolar disorder were contained rather than treated in the asylums of the 1800s. Such asylums were loosely transformed into psychiatric wards of government funded state hospitals, which were overwhelmingly large in size and low in quality as they housed "patients" unfit for the treatments offered or unmatched by the scientific and psychiatric knowledge of the time.

What is now classified as mental retardation, for example, was once considered "lunatic" behavior for which clinical professionals such as psychiatrists and neurologists had no explanation. DSM revisions steadily decreased confusion and increased harmony amongst clinical professionals and the public at the expense of thousands of "patients" wrongfully institutionalized and maltreated throughout the nineteenth and twentieth centuries. The unknown and slightly strange behaviors caused fear among the public inciting efforts to control, restrain, and isolate these behaviors from the controllable and unrestrained "normal". Here we see the first paradox of psychiatry as a medical science and the emergence of social constructionism (SC) in psychiatry, a theory claiming that our knowledge of the world is socially constructed through social exchanges and interactions.

Social perception of normal behavior was the driving force behind the classification of abnormal behavior. The use of language contributed to the distinction between abnormal and normal. Simply put, shifts in diagnostic categories made society more comfortable because it did not have to face bizarre behaviors that we now classify as mental retardation, intellectual disabilities, schizophrenia, and bipolar disorder just to name a few. This paper analyzes the theory of social constructionism and its role in the development of the DSM, and consequently, the act of diagnostic labeling. Given the historical development of the manual and its contributing factors, it can be argued that the DSM and diagnostic labeling are socially constructed entities rather than ones evidenced by science and the medical model that they claim to follow.

The medical model in relation to psychiatry

To further analyze the DSM as a function of SC, it is necessary to first carefully define certain terminology and to introduce psychiatry as a branch of medicine that subsumes the medical model, a term coined by the late psychiatrist Dr. R.D. Laing, who wrote extensively on the experience of psychosis. Merriam-Webster Dictionary defines psychiatry as "a branch of medicine that deals with the science and practice of treating mental, emotional, or behavioral disorders especially as originating in endogenous causes or resulting from faulty interpersonal relationships" (2014). As a medical specialty, psychiatry is assumed to follow the medical model, which holds the concept of disease at its core. The word "disease" refers to the deviation of normal functioning expressed by objective signs (i.e. elevated temperature and increased heart rate) and a subjective experience of distress as described by the patient. Furthermore, disease subsumes the germ theory, which requires that a biological component be present for a set of symptoms to be classified as a disease. In short, the medical model, even when pertaining to psychiatry, depends heavily on physical matter of the body. It does not account for the mind, an agency completely separate from the body, which lacks the physical matter that can be studied under a microscope.

In the late 1800s, terms such as "madness," "lunacy," or "insanity," once thought of as demonic possessions, were split into "mental diseases," which comprise modern day schizophrenia. Given the medical model, if mental illnesses such as depression, anxiety, or schizophrenia are to be considered "diseases," then they must have biological components. However, biological or genetic factors in such illnesses have yet to be scientifically proven. Therefore, the medical model of psychiatry, historically described as the "medical treatment of the soul" is in itself paradoxical and demonstrative of the lack of science in the DSM, since the soul cannot be treated medically. As already mentioned, the body is separate from the mind and mind is not matter. If mental illness cannot be supported by the medical model for the sole reason that the mind is separate from the body and lacks a biological component, then the workings of the mind must be built upon social constructionism.

Postmodern theories: Constructivism and Social Constructionism

Scientific modernist theories contend that truth can be discovered through objective scientific observation and measurement, or the use of the traditional scientific method. In contrast, there are several postmodern theories that challenge this by claiming, "there is not truth, only points of view" (Coady and Lehmann, 2008, p. 56.) Two of the more studied and accepted postmodern theories as applied to psychology are constructivism and SC. Constructivism holds that "there is no objective reality and that individuals mentally construct their own truths" (p. 56). Furthermore, SC focuses on the power of social interaction and culturally shared assumptions on the creation of knowledge, perception, and understanding; one's mental representations of reality are created through interpersonal social exchanges. The deeper meanings, or truth behind human behavior, are, in fact, language-based and socially constructed and have little to do with actual science. For the sake of this paper, constructivism and SC blend together to form the lens through which the DSM and diagnostic labeling will be analyzed.

If we revert back to the use of the medical model in relation to psychiatry, mental illness can be viewed as a socially constructed perception heavily influenced by language and linguistic context. Thomas Szacz coined the term "myth of mental illness" to suggest that the distinction between bodily illness and mental illness rests on a misuse of the term "illness" or "disease." If the term "disease" is restricted to observable biological or physiological abnormalities, than the term "mental illness" is a misnomer. Because mind is not matter, mental illness is simply a figure of speech. (Szacz, 2006, p. 43).

To further prove mental illness as a product of constructivism and SC, I explore the philosophical underpinnings of the relationship between language and emotion. In his article *The Social Constructionist Movement in Modern Psychology*, Gergen observes "many classic problems both in psychology and philosophy to be products of linguistic entanglement" and suggests that only with clarity concerning the nature and functions of the language will the problems be decomposed (1985, p. 267). Words like "sadness," "anger," and "happiness" are descriptions of our internal psychodynamic processes. Because emotions are experiences rather than objects of study, they acquire their meaning not from real-world referents but from the context of their usage. The interpretation of words given their context is also socially constructed as it depends heavily on the delivery of the experiences of the patient by the patient and the past experiences of the therapist.

Mental illness as a subjective experience

In patient-therapist relationships, language and communication form the foundation of assessment, diagnoses, and treatment interventions. Therapists assign diagnoses based on their perception of the patient's subjective experience, which, according to social constructionist views, may or may not be an accurate reflection of the patient's internal psychodynamics. In this line of thinking, accuracy of clinical judgment unnervingly rests on the ambiguity and subjectivity of linguistic composition. Hard scientific facts, unless supported by physiological indicators, are often absent in the assessment process that generates diagnostic labels. Therefore, given the lack of hard scientific knowledge to be observed during assessment processes for mental disorders, the DSM must also lack a significant amount of science, proving psychiatry to be more a construction of social interaction and linguistic composition rather than a branch of medicine in its traditional sense.

In a study conducted to distinguish between genuine and illusory mental health, clinical judgments by professionals were compared against various self-report scales by patients. Self-reports in this study consisted of hour-long verbal interviews between patient and clinician, and standardized written scales assessing emotional health. In *The Illusion of Mental Health*, Shedler et al. (1993) hypothesized that there are two subgroups among individuals who appear healthy on mental health scales: those that are psychologically healthy and those that are psychologically distressed. Those that are psychologically distressed maintain an illusion of mental health by appearing healthy on a scale but distressed per clinical judgment. Regardless of the direction of a patient's mental health, here lies the discrepancy between the patient's subjective experience and the clinical professional's objective interpretation.

In this study, clinicians evaluated each subject with respect to mental health-distress by using the Early Memory Test (EMT) as a basis for clinical inferences. In interpreting the EMT, clinicians "attended to qualitative factors such as how the self was represented, how the interpersonal world was represented, the affective tone of the material, and whether the memories were narratively coherent or contained in contradictions" (Shedler et al., 1993, p. 1121). Essentially, clinicians were attending to the linguistic composition and arriving at clinical judgments from a social exchange or interaction between themselves and the patients. Of the 287 subjects tested, 29 were judged distressed, 12 were judged healthy, and 17 were left unclassified because the clinical "data," or subjective experiences as told by the patients, were inconclusive and too sparse for analysis. Furthermore, an analysis of neurotic self-report measures and clinical evaluation in the same study demonstrated an asymmetric pattern. Individuals who reported distress were generally classified by the clinician as distressed. In contrast, subjects who reported health on the neuroticism self-report were "sometimes classified as distressed and sometimes classified as healthy" (Shedler et al., 1993, p. 1124).

To accurately assess a patient and assign a correct diagnosis, clinicians must overcome two barriers. The first barrier would be the discrepancy between the patient's overt emotional state and the patient's covert expressions. Often, patients report psychological health as a defensive denial of psychological distress. In other words, patients pretend to be psychologically healthy for the sake of their reputation while actually experiencing internal distress. They would rather hide their psychological distress than admit to having a mental health issue. The second barrier is the communication conveyed by the patient as evaluated by narrative believability. According to Shedler et al., narrative believability is the extent to which a patient's details of his or her subjective experience matches the overall narrative. For example, although a patient may recall his parent as warm and comforting, the overall narrative indicates an abusive relationship. Both barriers, whether conscious or unconscious to the patient, place a greater burden on the clinician or make it nearly impossible to accurately interpret the patient's true subjective experience. In which case, Shedler et al., informs the clinician to "trust your subjective impressions rather than the subject's explicit statements." Suddenly, the basis of interpretation of the patient's subjective experience no longer lies within the patient, but instead within the clinician's subjective impressions, which are shaped by his own experiences. The social construct of the clinician's life greatly influences the social construct of his patient's life and future.

DSM as a means of gaining social control

The DSM was created to differentiate between normal and abnormal. It was created to categorize not only disturbing pathological behavior, but behavior that was inexplicable and unaccounted for. In an attempt to distinguish between normal and abnormal for over a century, society has reached far beyond the tipping point. The most recent version of the DSM has a total of 947 pages and over 300 "disorders" ranging from severe clinical depression to acute stress. With over 300 disorders to choose from, the DSM appears to be more of a mental health shopping catalog rather than a tool for forming accurate clinical diagnoses. The ability to feel loss and sadness in order to experience happiness, for example, is no longer tolerated. It is labeled pathological and diseased. Unfortunately, society has begun to devalue the human's innate process of feeling and capacity to experience true, powerful emotions.

The flexibility of the DSM as measured by its hundreds of possible mental disorders has resulted in its misuse and overdiagnosis of its diagnostic categories. Depression, for example, has experienced lower and lower thresholds for clinical diagnosis as the DSM has evolved. Half a century ago, depression was either neurotic or melancholic. Now, the DSM includes every level of severity imaginable, from dysthymia, a form of mild depression lasting for two weeks at a time, to major depressive disorder, a debilitating state of profound sadness. In other words, all forms of natural human sadness are covered by a two hundred dollar book that allows for a lifetime supply of insurance coverage for medical treatments. Frankly, the only thing "medical" about psychiatry is the technology that has led to advancements in medications that now treat or minimize the symptoms of various mental disorders. It is this medication that has turned basic human emotions and life experiences into "diseases."

With catchall diagnoses at our fingertips, millions of people are spending money on various antidepressants, antipsychotics, and anti-anxiety pills that fuel the multi-billion-dollar pharmaceutical industry. Dr. Allen Frances, chair of the DSM-IV task force and author of *Saving Normal* and "Where Have All the Normals Gone?" states, "Grief is now Major Depressive Disorder; medical illness is Somatic Symptom Disorder; everyday worries are Generalized Anxiety Disorder; the forgetting of old age is Mild Neurocognitive Disorder; being geeky smart makes you an Aspie; gorging is Binge Eating Disorder; having temper tantrums is Childhood Bipolar Disorder; and all of us have Attention Deficit Disorder" (Frances, 2013). The DSM not only labels human behavior and emotion, but also persuades and manipulates its patients. I am almost tempted to replace the word "patients" with "customers" of America's booming and unnecessary pill-popping frenzy. However, despite the DSM's classification system of over 300 disorders, is the DSM itself really to blame for diagnostic hyperinflation?

The answer is no. From a social constructivist perspective, it can be argued that society is to blame for the expansion of abnormal psychiatry and the consequent reduction of normal. In an attempt to save "normal" by classifying what is abnormal, society has instead redefined normal, if such a thing ever existed. According to SC, "normal" was and is a construct of our interactions, experiences, and language. As a follower of a theory that invites one to challenge the objective basis of conventional knowledge and question what has been traditionally accepted, I have concluded that psychiatry and the DSM were built upon social constructions rather than having been medically derived. Society has redefined its perception and understanding of "normal" through various revisions of the DSM. What is considered normal today will only to be challenged yet again as societal knowledge, perceptions, and assumptions evolve in years to come.

References

American Psychiatric Association. (2000). *Diagnostic and statistical manual of mental disorders* (4th ed., text rev.). Washington, DC: Author.

American Psychiatric Association. (2013). *Diagnostic and statistical manual of mental disorders* (5th ed., text rev.). Washington, DC: Author.

Coady, N., & Lehmann, P. (Eds.). (2008). *Theoretical perspectives for direct social work practice*. 2nd ed. New York, NY: Springer.

Frances, D. A. (2013). *Saving Normal*. New York City: HarperCollins Publishers.

Frances, A. (2013, March 20). Where Have All the Normals Gone?. *TedWeekends*. Retrieved April 1, 2014, from http://www.huffingtonpost.com/allen-frances/jon-ronson-ted-talk_b_2978686.html.

Gergen, K. (1985). The Social Constructionist Movement in Modern Psychology. *American Psychologist* , *40*, 266-275.

Szasz, T. (2006). The Pretense of Psychology as Science: The Myth of Mental Illness in Statu Nascendi. *Current Psychology*, *25*, 42-49.

Feminist Critique: Posttraumatic Stress Disorder

Kaci Woodcock

Introduction

Post Traumatic Stress Disorder or PTSD has been talked about a lot since the United States sent military troops into Iraq and Afghanistan, but military personnel are not the only people affected by PTSD. Approximately eight percent of American adults will experience PTSD in their lifetimes; for military veterans the rate is closer to eleven percent (Boone K. N., 2011). Highest rates are found among survivors of rape, combat, captivity and ethnically or politically motivated internment or genocide (American Psychiatric Association, 2013). With the DSM-5 came changes to the structure and groupings of diagnoses, and as the standard of diagnosis in the US, it is important to continually critique and analyze these diagnoses as there are no biologically based tests to prove mental health diagnoses. There are also many theories with which to view and understand mental health, which affect both diagnosis and treatment. The purpose of this chapter is to analyze and critique the DSM-5 diagnosis of Post Traumatic Stress Disorder from a feminist perspective. The history of PTSD as a diagnosis, the current PTSD criteria, along with the recent changes to the criteria, feminist theory, gender issues related to PTSD, and an analysis will be covered through this chapter.

History

The diagnosis of PTSD was first introduced in the *Diagnostic and Statistical Manual of Mental Disorders* (DSM III) that was published in 1980. The diagnosis was a result of political advocacy mainly on the part of Vietnam War veterans. The veterans wanted the diagnosis created to reduce the stigma of having prolonged psychological symptoms and distress after experiencing combat. But the veterans also had support from clinicians who had studied other forms of trauma, such as rape, and Holocaust survivors (Boone, 2011).

As a diagnosis, PTSD stems from the military. PTSD was originally called shell shock, which was believed to be psychological distress caused by concussions from the new high explosive weapons technology. Even though the understanding of what was causing the psychological distress was quickly seen as invalid, the term remained (Boone, 2011). Then during WWII the US department of Veterans Affairs (VA) was overwhelmed with patients with psychological stress. The VA then came up with the term "war neurosis," and they believed it to be a pre-existing condition from inherent weakness or defective parenting that was exacerbated by combat (Boone, 2011). After that people who were diagnosed with "war neurosis" were immediately discharged. This term and understanding also quickly faded out mainly due to lack of manpower. The diagnosis changed once again from "war neurosis" to "combat fatigue" and went from being pathological to a common/normal reaction to stress (Boone, 2011).

Psychiatrists from the VA theorized that the best treatment path was rest, emotional support, and encouragement not to think of themselves as sick or abnormal. Under this treatment plan, most men were able to return to combat fairly quickly (Boone, 2011). One problem, however, was that the term "combat fatigue" was not transferable to the general public; however, there were cases of civilians experiencing similar psychological distress from experiencing trauma (Boone, 2011). To alleviate this problem the first DSM included "gross stress reaction" as a way to cover the general public's reaction to stress, but it was taken out in the second DSM (Boone, 2011). PTSD in DSM-III had the criteria that the disorder stemmed from a precipitating external event (Boone, 2011). Traumatic stress was defined as "a recognizable stressor that would evoke significant symptoms of distress in almost everyone; qualifying events fell generally outside the range of usual human experience." (Boone, 2011).

Current Criteria

The current DSM-5 diagnostic criteria include exposure to trauma, presence of intrusive symptoms, avoidance of stimuli, negative alterations in cognition or moods, changes in reactivity and arousal, and disturbances causing distress in social, occupational or other areas of functioning. These disturbances are not due to the effects of substances. Intrusive symptoms include recurrent and involuntary distressing memories of the event, distressing dreams that either contain content or the affect of the dream is related to the traumatic event, flashbacks where the person feels the event is reoccurring, and reactions to external or internal stimuli related to the trauma. One or more of the intrusive symptoms must be present. Some of the negative alterations in cognition or mood symptoms are inability to remember aspects of the trauma, persistent and exaggerated negative beliefs about oneself or the world, blaming of oneself for the trauma, diminished interest and participation in activities, feelings of isolation, and difficulty experiencing pleasure. Two or more negative alterations in cognition or mood symptoms must be present for a diagnosis of PTSD. Also the symptoms must last for more than one month (American Psychiatric Association, 2013).

Changes to the diagnosis of PTSD from the DSM-IV-TR

The DSM-5 differs from the DSM-IV-TR in the diagnosis of PTSD. For example, the DSM-5 is more explicit regarding how the individual experienced the trauma. It specifies if the trauma was directly experienced, witnessed, or secondary survivor (i.e. primary survivor is a family member or friend) (Hiler, 2013). Also, the subjective reaction, which was criterion A2 in DSM-IV-TR, has been removed. This criterion stated, "The person's response involved intense fear, helplessness or horror" (Hiler, 2013). In the DSM-IV-TR arousal and numbing symptoms were listed under the same criteria cluster; however, the DSM-5 split the arousal and numbing symptom cluster from one into two criteria. The first is avoidance and the other is persistent negative alterations to cognition and mood. Under the criterion of persistent negative alterations to cognition and mood most of the numbing symptoms are retained and new or re-conceptualized symptoms such as persistent negative emotional state were added (Hiler, 2013). The DSM-5 is also more developmentally minded, as changes to the age criteria have been made. The age has been lowered to six and now there are separate criteria for children under the age of six (Hiler, 2013).

Feminist Theory

Feminist theory includes a wide variety of theories that differ in focus and population (Saulnier, 255). However, there are some basic principles that can be applied to most versions of feminist theory.

Feminist theory is empowerment based and is a holistic view of the interrelationship between social, political, spiritual, intellectual and material aspects of human existence (Robbins, Chatterjee, & Canda, 2006). The theory is focused on the need to analyze and deconstruct the discriminatory aspects of ones existence, mainly the discriminatory aspect of patriarchy (Robbins, Chatterjee, & Canda, 2006). The theory posits a need to name the contributing factors to the discrimination and oppression of minority groups, such as expectations, language, social arrangements and behaviors (Robbins, Chatterjee, & Canda, 2006).

There are many strengths and weaknesses of feminist theory. One strength is that the theory aids in explaining gender-based and other attributed identity-based disparities (Saulnier, 269). The theory acts as a guide for practitioners in efforts to take apart the structural and interpersonal limits and constrictions a client faces (Saulnier, 269). Feminist theory provides a lens to analyze the various forms and layers of oppression faced by women and other minority groups (Saulnier, 269). However, some argue that some branches of feminism are too narrow, which limits the populations and people to whom the theory can be applied (Saulnier, 269). Also, there is a lack of political viability of the more comprehensive theories because so many people and populations, typically dominant populations, are left out (Saulnier, 269).

Gender issues and PTSD

PTSD is more prevalent among females than males, and women experience PTSD for longer time periods than do males (American Psychiatric Association, 2013). Potentially, these discrepancies in prevalence and duration are due to women being more likely to experience traumatic events, such as rape and interpersonal violence. (American Psychiatric Association, 2013). There have been a few studies done in an effort to understand the differences between the genders when it comes to PTSD. One study done by Norris, Perilla, Ibañez, & Murphy (January 01, 2001) studied the effect of traditional gender roles on the higher rate of PTSD in women than men. The study was a cross-cultural study between Mexico and the United States and studied PTSD in women and men after they had experienced similar natural disasters. The natural disasters were Hurricane Andrew, which occurred in South Miami in August 1992 and Hurricane Paulina which occurred in Acapulco, Mexico and surrounding areas in October 1997. The authors analyzed the difference in PTSD symptoms and rates between men and women following the disasters across three main cultures, Non-Hispanic white U.S. natives, African-American U.S natives, and Latino Mexican natives. The authors rated Latino Mexicans as the most "traditional" in gender roles and African-American US natives the least "traditional" and believed that there would be higher rates and effects of PTSD on women in the culture with more "traditional" gender roles. The study found Mexican women were the most highly distressed compared to their male counterparts, and US non-Hispanic white women were only somewhat more distressed than their male counterparts (Norris, Perilla, Ibañez, & Murphy, 2001). African-American U.S natives showed the least amount of distress when compared to their male counterparts and to the other two groups of women (Norris, Perilla, Ibañez, & Murphy, 2001).

Studies have also been done to see if the difference in gender was due to a difference in social support. A study by Andrews, Brewin, & Rose, (August 01, 2003) predicted that the difference in rates of PTSD among the genders is actually due to the differences in social support, both positive and negative, and the perceptions of social support received. They found that women reported more negative responses from family and friends following trauma than men, which had a correlation to higher rates of PTSD symptoms. The impact of both positive and negative social support was greater for women than for men (Andrews, Brewin, & Rose, August 01, 2003).

Social defeat, a term used in Britain to describe the sense that one has failed struggles or is losing social rank, was also studied as it pertains to gender differences in PTSD (Troop, & Hiskey, (November 01, 2013). The purpose of the study was to determine if constructs related to social rank/defeat affect the predictability of the diagnosis of PTSD. The authors used self report measures on PTSD symptoms (Post traumatic diagnostic scale) and on social rank which included: The Social Comparison Rating Scale (SCRS), The Social Defeat Scale (SDS), and The Internal–External Entrapment Scale (IEE-EXT measures the perception of things in the outside world or IEE-INT internal feelings and thoughts). The authors found social defeat, but not other aspects of social rank, was predictive of a diagnosis of PTSD. Higher levels of social defeat at baseline predicted less improvement in PTSD symptoms over time and reduction in social defeat over time predicted greater improvement in PTSD symptoms (Troop, & Hiskey, (November 01, 2013).

Analysis and Critique

As seen in the previous section, there is a difference between men and women when it comes to rates of PTSD and likelihood of its development. These differences have been studied, and attempts to discover the causes and explanations for the differences have been made. However, traditionally, psychological theories were developed based on studies done with men and then generalized to women. At this time, there are still no concrete answers as to why there are differences between men and women. Since feminists' main focus is on naming the barriers and discriminations women have faced, they would attribute these differences to those discriminations. One example is traditional gender roles as seen in the Norris, Perilla, Ibañez, & Murphy, (January 01, 2001) cross-cultural study. Other issues that face women are social supports as seen in the study conducted by Andrews, Brewin, and Rose, (August 01, 2003). When women face more negative social supports and victim blaming, there is a higher chance of the development of PTSD. Women are also more affected by social supports than men. The study suggested that this could be due to women participating in more social interaction than men (Andrews, Brewin, & Rose, August 01, 2003). The DSM-5 diagnosis of PTSD does not take into account any of the social, political, economic and other factors in PTSD diagnosis even though all of those factors seem to have an effect on PTSD and have a higher impact on women.

Using the DSM leads mental health practitioners to follow the medical model of health. This is a problem according to feminist theorists, as little effort is made to understand how the "lifeworld" and the identity of the individual and those around her have been altered due to the trauma (Ballou, & Brown, 2002). Instead. the focus is on symptom removal. Symptom removal is even medicalized and standardized across the types of trauma (Ballou, & Brown, 2002). If gender-linked violence issues are viewed from a medical lens, the sociopolitical nature of the violence is ignored or negated. As noted by Ballou and Brown, "If diverse instances of violence, atrocity, betrayal and sexual invasion are lumped together as 'trauma' their psychic meaning, interpersonal significance, and moral implication are lost to view" (Ballou, & Brown, 2002).

According to feminist theorists, a PTSD diagnosis can have attached meanings that affect the survivor (Ballou, & Brown, 2002). According to feminist theories/theorists, having the diagnosis of PTSD will mean that the victim or survivor will inevitably and uniformly suffer lasting or permanent psychological damage (Ballou, & Brown, 2002). While that could be true for some survivors, other survivors might not suffer lasting or permanent psychological damage. This again emphasizes the point that the way the diagnosis is written and understood takes away from recognizing the individual and his or her personal circumstances. PTSD is seen as a part of a mental health caste system, where women with PTSD are seen as "good normal girls" responding to an abnormal situation, whereas women with Borderline personality disorder are seen as "bad girls" (Ballou, & Brown, 2002). This relates back to the effects of social supports and stigma on PTSD development.

The original wording of the diagnosis implied that the only reaction for almost everyone who experienced a traumatic event was to develop PTSD (Boone, 2011). "If you react normally to trauma, you have a disorder; if you react abnormally, you don't." (Boone, 2011). The symptoms of PTSD are fairly normal and common. A study done by Harvard psychiatrist J. Alexander Bodkin found people experience symptoms after events that do not reach the threshold of Criterion A. Some of those events were divorce, the collapse of adoption arrangements, and financial ruin. (Boone, 2011). Another study by psychologists at Temple University took a set of "normal" undergraduate students, and when asked to think about most troubling aspect of their lives, the students reported experiencing symptoms of PTSD at equal or higher rates than student exposed to Criterion A traumas (Boone, 2011). If "normal" traumas can cause PTSD symptoms then there is a problem with the criteria, as there can be little distinction between those dealing with normal distress and people who truly suffer from PTSD (Boone, 2011). While there is a problem with the criteria as shown by the fact that people exhibit symptoms of PTSD without having experienced a trauma that is included under criterion A, the symptoms are real and distressing.

Conclusion

As a diagnosis, PTSD originated from the military, a historically male dominant institution, and the original terminology of "shell shock," "war neurosis," or "combat fatigue," was not transferable to the general public. The DSM mentions the fact that women experience PTSD more than men and some of the other cultural, social and economic factors that affect PTSD but these factors are not included in the criteria and are only used as a mention regarding prevalence. The DSM is a medicalized model of mental health, which detracts from the understanding of the individual's experience of the trauma and the symptoms. It disregards the sociopolitical environment of the victim, ignores the difference between gender violence and other interpersonal violence. Also it does not take into account the compounding effects of society such as economics, social structure, social defeat, family history, and previous trauma and how that can increase the intensity of PTSD symptoms.

While the diagnosis has many problems and discounts many of the sociopolitical factors that contribute to PTSD, the diagnosis can still serve a purpose and have a positive effect. While it has been shown that PTSD is a common and normal reaction to an abnormal event, the symptoms are still distressing for the individual. Since affected individuals still experience distress, intrusive symptoms, and decreased functionality in life, PTSD should still be a diagnosis in the DSM.

However, when it comes to applying the diagnosis it should be done in a way that builds up the survivor and does not stigmatize him or her further. It should be emphasized that PTSD is a normal and common reaction to stressful life events, especially criterion A traumatic events. When diagnosing PTSD, it should be noted that it does not mean that the individual is reacting abnormally or is weaker than someone who does not develop PTSD. Even though PTSD is a common reaction to experiencing trauma or even distressing life events, not everyone exposed to such experiences develops PTSD or seeks treatment and thus is included in the statistics.

Each individual will experience trauma differently, and each person comes to the experience with a different history and background. Thus, the treatment of each individual needs to be multifaceted, taking into account his or her gender background, socioeconomic background, cultural background and trauma history. While the DSM-5 results in mental health practitoners following the medical model of diagnosis, treatment should not follow suit and become medicalized.

References

American Psychiatric Association. (2013).*Diagnostic and Statistical Manual of Mental Disorders: DSM-5*. Arlington, VA: American Psychiatric Association.

Andrews, B., Brewin, C. R., & Rose, S. (August 01, 2003). Gender, Social Support, and PTSD in Victims of Violent Crime. *Journal of Traumatic Stress, 16,* 4, 421-427.

Ballou, M. B., & Brown, L. S. (2002). *Rethinking Mental Health and Disorder: Feminist Perspectives*. New York, N.Y: Guilford Press.

Boone, K. N. (January 01, 2011). The Paradox of PTSD: The Tortured History of Posttraumatic Stress Disorder as a Psychiatric Diagnosis Reveals a Great Deal about Why the Military is Struggling to Treat American Soldiers Who Bear Deep Psychological Scars. *Wilson Quarterly, 35, 4, 18-22.*

Hiler, A. (June, 6, 2013). *Highlights of Changes from DSM-IV-TR to DSM-5.* Retrieved from http://www.dsm5.org/Documents/changes%20from%20dsm-iv-tr%20to%20dsm-5.pdf

Norris, F. H., Perilla, J. L., Ibañez, G. E., & Murphy, A. D. (January 01, 2001). Sex Differences in Symptoms of Posttraumatic Stress: Does Culture Play a Role?. *Journal of Traumatic Stress, 14,* 1, 7-28.

Robbins, S. P., Chatterjee, P., & Canda, E. R. (2006). *Contemporary Human Behavior Theory: A Critical Perspective for Social Work.* Boston: Allyn and Bacon.

Saulnier, C. F. (2001). Feminist Theories. In Lehmann, P. Editor & Coady, N. Editor (Eds.), *Theoretical Perspectives for Direct Social Work Practice: A Generalist-Eclectic Approach* (255-275). New York, NY: Springer Publishing Company

Troop, N. A., & Hiskey, S. (November 01, 2013). Social Defeat and PTSD Symptoms Following Trauma. *British Journal of Clinical Psychology, 52,* 4, 365-379

Gender Dysphoria through a Feminist Lens

Andrew Zapke

Introduction

The American Psychological Association first introduced the concept of Gender Identity Disorder in 1980 with the DSM-III edition. Since then, the concept and diagnosis have undergone significant changes with the rising awareness of transgender issues. The Gender Dysphoria diagnosis is widely debated by activists and psychologists, in the popular media and in academic literature. Contemporary frameworks embed a critical analysis of the Gender Dysphoria diagnosis in feminist theory.

Gender is the result of a process of socialization; gender roles are ever-shifting. While feminist theory has long distinguished between gender assigned at birth (natal gender) and expressed gender, and the process by which society shapes gender roles, society has been slower to acknowledge these nuances. Feminist theory, as shown in this paper, offers valuable concepts for such a critical integration.

While the inclusion of the Gender Dysphoria diagnosis in the DSM has been lauded by some practitioners for its usefulness in securing treatment such as Gender Reassignment Surgery and psychological treatment, the effect of patriarchy on transgender individuals and the distress caused by carrying a mental health diagnosis should not be ignored.

Feminism

Feminism has a long history and many branches. We will interpret and critique psychoanalysis, and in particular, Gender Dysphoria, using Radical Feminism.

Feminist theory is useful for analyzing the experiences of people from a gender-based perspective. It can help explain the social and environmental factors that create or contribute to problems encountered by groups and individuals (Saulnier, 2008). Feminism, in comparison, can be used to describe a political, cultural or economic movement aimed at establishing equal rights and legal protection for women. Feminism draws from and informs theories concerned with issues of gender difference, as well as being a movement that advocates gender equality for women and campaigns for women's rights and interests.

Feminist analysis is about explicitly putting women back into the historical record, where they have always been but were never honored or identified. This de-emphasis on the historical roles of women contradicts models of analysis that call for pure objectivity and value neutrality in the pursuit of knowledge. (Kanenberg, 2013)

Feminist theory, with its "holistic view of the interrelationships between material, social, intellectual, and spiritual facets of human existence" (Robbins, Chatterjee, & Canda, 2006, p. 97), is useful for examining social policies. It is also an appropriate method for looking at these factors and their impact on individuals and communities. Feminist theory asks how gender is constructed and examines how social policies on several levels are used to enforce gender relations (Kanenberg, 2013).

Diagnosis

The Diagnostic and Statistical Manual of Mental Disorders, 5th ed. [DSM](American Psychiatric Association, 2013)*,* addresses the issue of gender variance in a stand-alone chapter entitled "Gender Dysphoria."

In the previous edition of the DSM, the DSM-IV, Gender Identity Disorder focused on the issue of identity, specifically the incongruity between someone's natal gender and the gender with which he or she identifies. While this incongruity is still crucial to the DSM-5 diagnosis of Gender Dysphoria, there is now an emphasis on the distress or social impairment associated with the condition. This shift recognizes, in theory, that the disagreement between natal gender and gender identity may not necessarily be pathological if it does not cause distress. The diagnosis hinges on distress regarding the incongruence between natal and expressed genders.

The DSM-5 now distinguishes between Gender Dysphoria in children and Gender Dysphoria in adolescents and adults; for the rest of this paper, when referring to Gender Dysphoria, we will be referring to Gender Dysphoria in Adolescents and Adults. Gender Dysphoria can be defined as follows:

> Gender Dysphoria refers to the distress that may accompany the incongruence between one's experienced or expressed gender and one's assigned gender. Although not all individuals will experience distress as a result of such incongruence, many are distressed if the desired physical interventions by means of hormones and/or surgery are not available. (APA, p.451)

The core of the Gender Dysphoria diagnosis is the incongruence between natal gender and expressed gender. Non-binary experienced gender identities and expression are acknowledged in the diagnostic criteria. This allows for the expressed gender to be any experienced gender differing from the natal gender, not simply the "opposite" gender. This allows gender to exist along a spectrum. Given the increased openness of gender expressions under this definition,   it is important that the diagnosis be limited to "those individuals whose distress and impairment meet the specified criteria" (APA, p. 458).

Initial Critique of Diagnosis

The DSM focuses on distress as the clinical problem, not identity per se. However, the distress required for Gender Dysphoria is where the diagnostic usefulness falls apart.

Gender Dysphoria is marked by "clinically significant distress or impairment in social, occupational or other important areas of functioning." (APA, p. 453) 'Clinically significant distress' is also a diagnostic criterion for PTSD, anxiety and a number of other disorders listed in the DSM. Therefore, the only clinical factor that makes Gender Dysphoria a mental disorder could be attributed to almost any other mental disorder listed.

Distress is not an inherent part of being transgender. This sets it apart from many other disorders in the DSM. The distress that may accompany Gender Dysphoria arises, not from the incongruence, but as a result of a culture that stigmatizes people who do not conform to gender norms. For instance, many transgender people are not distressed by their experienced gender and should not be diagnosed with Gender Dysphoria. Hence, the updates to the Gender Dysphoria diagnosis continue to be a particularly interesting part of the DSM, and social workers need to be both current and discerning about these updates..

Implied in such a diagnosis is a belief that what ties transgender people together is their mutual disdain for certain parts of their bodies (usually assumed to be genitals). Not all individuals who are diagnosed with Gender Dysphoria, however, wish to alter their body through hormone therapy or surgery. There are transgender people who do not feel emotional pain caused by their bodies, but rather connect with and love their bodies. Requiring distress is harmful; it implies that for someone to be the gender they are, they need to want the stereotypical body parts of that gender. Requiring distress implies women cannot happily have penises and men cannot happily have vaginas.

Langer and Martin (2004) question the therapeutic purpose of Gender Identity Disorder, but the objections they raise are still relevant under the Gender Dysphoria diagnosis. In particular, they discuss ethical objections to the rationale for treatment. This comes out nicely in the following quotation outlining this criticism; Langer and Martin (2004), quote Wilson:

> The psychiatric interpretation of inherent transgender pathology serves to attribute the consequences of prejudice to its victims, neglecting the true cause of distress. It promotes treatment paradigms that are punitive rather than affirmative with the goal of conformity and not self-acceptance.

Additionally, the DSM has a hard time taking into account the difference between inherent distress and distress that is socially imposed (Langer and Martin, 2004).

Culture of (Trans) Misogyny

Most societies allow for the existence of only two genders. To maintain this gender binary, most cultures traditionally insisted that every individual be assigned to the category of either man or woman at birth, and that individuals conform to the category to which they have been assigned (Drescher, 2010). The categories of "man" and "woman" are considered to be mutually exclusive.

Radical feminism, which approaches the injustices faced by individuals as political issues grounded in power imbalances, is a useful method to examine the issue of Gender Dysphoria as a social issue rather than a mental health issue. It succeeds with this perspective because it characterizes society as patriarchal. Society is constructed in such a way that men possess a disproportionate share of power (Saulnier, 2008). All human interactions are affected by this male privilege. The core cultural ideas of what is normal and valued are associated with masculinity (Robbins, p. 117). Rigid gender roles lead to oppression, which acts as the main tool to maintain the system of patriarchy.

Society, as a patriarchal, male-dominated institution, values masculinity over other gender expressions. Gender roles are well-defined and rigorously defended. People who violate those gender roles are seen as deviant. Despite increased awareness in society, transgender people face significant discrimination and disproportionate violence while navigating a system that is based on a binary notion of sex and gender (Taylor, 2007).

Joan Acker (2006) contends that, "masculinities are essential components of the ongoing male project, capitalism" (p. 129). Our society, which is based on capitalism, values masculinity. Being a man involves cultural images and practices (Acker, 2006). Masculinity "enshrines particular male bodies." (Acker, 2006, p. 131)

Masculinity and violence are connected in this system. Masculinity is often organized around domination and control. As Robbins, Kimmel and Messner note, gender, class, and race are the central methods through which power is expressed. We can draw from articles about masculinity more generally as we look at this question of how power and resources are distributed in society (2004, p. ix).

For example, Vandello et al. (2008), explore the tenuous nature of manhood. Men may be especially sensitive to the precariousness of their social status. When presented with feedback that they did not measure up to others of their gender in a gender identity test, men showed increased anxiety, heightened desire to hide this feedback, and a defensive insistence that they would "do better" on future tests. Given that manhood is precarious, requiring action and success in all "manly" endeavors, it is not surprising that many men feel anxiety over what they perceive as an unattainable standard. (Vandello, et al., 2008)

This anxiety causes men (as the ultimate arbiters of societal norms) to be wary of allowing other gender expressions in their "men's only club." Women occupy only 3.8% of Fortune 500 chief executive officer seats and represent only 3.2% of the heads of boards in the largest companies of the European Union. In 2012, women held only 90 of the 535 seats (16.8%) in the U.S. Congress and 19.1% of parliamentary seats globally (Paustian-Underdahl, Walker, & Woehr, 2014). Male privilege works to lock out women and gender non-binary individuals from fully participating positions of power.

Kroeper, Sanchez, and Himmelstein, (2014), expand the view of precarious masculinity by showing that masculinity requires social proof. They explored the use of prejudice against sexual minorities to affirm masculinity. In their view, masculinity must be consistently reaffirmed in order to be maintained. (Kroeper, Sanchez, and Himmelstein, 2014).

Transgender women, especially, in transitioning from male to female, give up access to male privilege. To do so in a society that prizes masculinity over all else is unforgivable. Haraldsen and Dahl (2000) found that older research into Gender Dysphoria shows that men who wanted to become women were considered to have more severe psychopathology than women wishing to become men, all with no empirical evidence to support this claim.

This leads to transphobia. We can find an example of this transphobia, specifically transmisogyny, in the article by Kevin Williamson for the National Review Online. In it, Williamson argues that Laverne Cox, a transgender actor, is "an effigy of a woman." Williamson constantly uses the wrong gender pronoun while referring to Cox, and makes several references to delusion and genital mutilation. It is no wonder that a transgender person might experience "clinically significant distress" (APA, p. 453) in a culture in which such articles are published.

It is understandable that many transgender individuals, already stigmatized for their expressions of gender variance, would wish to avoid the added burden of being labeled as having a mental disorder. This is especially true for members of the transgender community who are not anatomically dysphoric and who neither seek nor desire medical or surgical intervention to change their bodies. Further, many in the transgender community who do seek medical intervention prefer being diagnosed with a medical condition, rather than a psychiatric disorder. (Drescher, 2010)

Social norms

People express gender beliefs, their own and those of the culture in which they live, in everyday language as they accept and assign gendered meanings to what they and others do, think, and feel (Drescher, 2010). Gender beliefs touch upon almost every aspect of daily life.

Likewise, social norms are part of our everyday life. They help people self-organize in many situations; the responsibility to enforce social norms is not the task of a central authority but a task of each member of the society (Horne, 2009). Research shows that a key reason people enforce norms is their social relations (Horne, 2009). People care about what others think of them because they want to maintain valued relationships.

Social norms are part of the inertia that keeps Gender Dysphoria a diagnosed mental health issue. Psychologists are driven to maintain their relationships with others in the healthcare industry, so they do not push for a change to the DSM. Just as the APA was quick to affirm that they did not endorse homosexuality despite dropping it from the DSM, so too the APA does not want to be seen endorsing the politically hot button topic of transgender identity.

Politics has always been connected to psychiatric diagnosis. It was once proposed that rape be included as a sexual mental disorder; there were protests that the new diagnosis would be used as a mitigating excuse by a rapist at his trial. (Silverstein, 2009)

Schilt and Westbrook (2009) explore the process by which transgender men, as new recipients of male privilege, get accepted in the workplace. Their studies show that gender roles are still firmly reinforced. As soon as a transgender man transitions to presenting as a man in the workplace, he is expected to lift, push and move things far more frequently than the cisgender women in the office. This underlines the point that feminism is inextricably linked to any critical analysis of transgender politics

Psychiatric diagnosis

Although research results (Haraldsen and Dahl, 2000) have shown that transgender people are no more likely to suffer from psychopathology than cisgender people, providers may perceive unrelated mental health issues to stem from a person's gender identity. Clinically referred adults with Gender Dysphoria may have coexisting mental health problems, most commonly anxiety and depressive disorders. (DSM, p. 459)

Providers may also assume that a person's gender identity stems from a mental health issue. This presents a situation in which it can become difficult for a transgender person to obtain appropriate care for mental health issues. Providers rarely receive information presented in a way that allows them to see a person's mental health issues as separate from transgender status. This may result in inadequate or inappropriate care. (Bauer, et al, 2009) Hepp, Kraemer, Schnyder, Miller and Delsignore, (2005) found that lifetime psychiatric comorbidity is high for individuals with Gender Dysphoria.

The DSM acknowledges that Gender Dysphoria is associated with stigmatization, discrimination, and victimization. Additionally, access to health services and mental health services may be impeded by structural barriers, such as institutional discomfort or inexperience in working with this patient population. (APA, p. 458)

Cuevas, Finkelhor, Ormrod, and Turner, (2009) researched the relationship between victimization and lifetime psychiatric diagnosis. Psychiatric diagnoses typically have been emphasized as a consequence of victimization, with less research examining if it also functions as a risk factor for further victimization. Cuevas, Finkelhor, Ormrod, and Turner, (2009) have shown that children with a psychiatric diagnosis have significantly higher rates of victimization than children without a psychiatric diagnosis. The results highlight the need to consider psychiatric diagnoses as a risk marker for past and possible future victimization.

Directions forward

The Gender Dysphoria diagnosis might be seen as patronizing, demeaning and stigmatzing. There are parallels between Gender Dysphoria and homosexuality that should be considered. In DSM-II, published in 1968, homosexuality was reclassified as a "sexual deviation." However, by 1970, scientific research arguing for a non-pathological view of homosexuality was brought to the APA (Drescher, 2010).

The APA wrestled with the question of what constitutes a mental disorder. They eventually concluded that homosexuality per se was not a mental disorder because it did not cause "generalized impairment in social effectiveness of functioning" (Drescher, 2010. P. 434). Homosexuality was removed from the DSM in 1973 (Drescher, 2010).

There are parallels between the views on homosexuality and gender identity in psychiatric and medical thinking. Historically, transgender people have been "the most visible minority among people involved in same-sex sexual practices. As such, transgender people have been emblematic of homosexuality in the minds of most people" (Devor, 2002, p. 5). Just as homosexuality did not cause impairment enough to be classified as a mental disorder, neither does Gender Dysphoria create diminished social functioning.

Lev (2013) points out that in 2012, the APA itself released a statement affirming the necessity of surgical and hormonal care, as well as civil rights protections, for transgender people.

Conclusion

"Social workers pursue social change, particularly with and on behalf of vulnerable and oppressed individuals and groups of people." (NASW Code of Ethics). Transgender individuals are certainly oppressed in today's patriarchal society.

From the perspective of social work theory and practice, and drawing from radical feminist theory, the inclusion of Gender Dysphoria in the DSM must be regarded as problematic and deserves continuing critical discussion. Listing Gender Dysphoria as a psychiatric disorder mainly reflects existing (though ever-changing) social attitudes and prejudices which serve as forms of social control. Locking these prejudices into diagnoses stands at odds with attempts to depathologize gender variation. If the DSM continues to list Gender Dysphoria as a disorder so as to provide insurance-covered treatment for gender-variant individuals, it should take steps to ensure that its description embraces the great multiplicity of gender variance, does justice to the whole range of conditions of health and well-being experienced by transgender persons, and acknowledges the historically changing social pressures that create and maintain health problems related to gender.

References

Acker, J. (2013) Is capitalism gendered and racialized? Pp. 125-133. In Andersen, M., & Collins, P. (Eds.). *Race, Class, and Gender. An anthology* (8th Ed.).Belmont, CA: Wadsworth Publishing.

American Psychiatric Association. (2013). *Diagnostic And Statistical Manual Of Mental Disorders* (5th ed.). Arlington, VA: American Psychiatric Publishing.

Bauer,G., Hammond, R., Travers, R., Kaay, M., Hohenadel, K., Boyce, M. (2009). "I don't think this is theoretical; this is our lives:" How erasure impacts health care for transgender people. *Journal of the Association of Nurses in AIDS Care*, Vol. 20, No. 5, September/October 2009, 348-361

Cuevas, C., Finkelhor, D., Ormrod, R., & Turner, H. (2009). Psychiatric Diagnosis as a Risk Marker for Victimization in a National Sample of Children. *Journal of Interpersonal Violence, 24,* 4, 636-652.

Daley, A., & Mulé, N. J. (2014). LGBTQs and the DSM-5: A Critical Queer Response. *Journal of Homosexuality, 61,* 9, 1288-312.

Devor, H. (2002). Who are "we"? Where sexual orientation meets gender identity. *Journal of Gay & Lesbian Psychotherapy,* 6(2), 5–21.

Drescher, J. (2010). Queer diagnoses: parallels and contrasts in the history of homosexuality, gender variance, and the diagnostic and statistical manual. *Archives of Sexual Behavior, 39,* 2, 427-60.

Haraldsen, I. & Dahl, A. (2000). Symptom profiles of gender dysphoric patients of transsexual type compared to patients with personality disorders and healthy adults. *Acta Psychiatrica Scandinavica,* 102, 276-281

Hepp,U., Kraemer, B., Schnyder, U., Miller, N., Delsignore, A. (2005). Psychiatric comorbidity in gender identity disorder. *Journal of Psychosomatic Research* 58 (2005) 259–261

Horne, C. (2009). A social norms approach to legitimacy. *American Behavioral Scientist, 53,* 3, 400-415.

Kanenberg, H. (2013) Feminist policy analysis: Expanding traditional social work methods. *Journal of Teaching in Social Work,* 33:2, 129-142

Kroeper, K., Sanchez, D., and Himmelstein, M. (2014), Heterosexual men's confrontation of sexual prejudice: The role of precarious manhood. *Sex Roles* 70:1–13

Langer, S., and Martin, J. (2004). How dresses can make you mentally ill: Examining gender identity disorder in children. *Child and Adolescent Social Work Journal,* 21, 1, 5-23.

Lev, A. I., (2013), Gender Dysphoria: Two steps forward, one step back. *Clinical Social Work Journal.* September 2013, Volume 41, Issue 3, pp 288-296

National Association of Social Workers. (2008) Code of Ethics. Retrieved July 24, 2014, from https://www.socialworkers.org/pubs/code/code.asp

Paustian-Underdahl, S. C., Walker, L. S., & Woehr, D. J. (2014). gender and perceptions of leadership effectiveness: A meta-analysis of contextual moderators. *Journal of Applied Psychology.*

Robbins, S. P., Chatterjee, P., & Canda, E. R. (2012). *Contemporary human behavior theory: A critical perspective for social work.* Boston: Allyn and Bacon.

Saulnier, C. F. (2008). Feminist theories, pp.343-368. In N. Coady, & P. Lehmann, (Eds.). *Theoretical perspectives for direct social work practice.* 2nd ed. New York, NY: Springer.

Schilt, K. and Westbrook, L. (2009). Doing gender, doing heteronormativity: "gender normals," transgender people, and the social maintenance of heterosexuality. *gender & Society* 2009 23: 440

Silverstein, C. (2009). The implications of removing homosexuality from the DSM as a mental disorder. *Archives of Sexual Behavior, 38,* 2, 161-3.

Taylor, J. (2007). transgender identities and public policy in the United States: The relevance for public administration. *Administration & Society.* Volume 39 Number 7. November 2007. 833-856

Vandello, J., Bosson, J., Cohen, D., Burnaford, R., and Weaver, J. (2008), Precarious manhood. *Journal of Personality and Social Psychology* Vol. 95, No. 6, 1325–1339

Williamson, K. (May 30, 2014), Lavern Cox is not a woman. *National Review Online.* Retrieved from http://www.nationalreview.com/article/379188/laverne-cox-not-woman-kevin-d-williamson